The Star of Bethlehem
A Skeptical View

By Aaron Adair

The Star of Bethlehem: A Skeptical View
Copyright © 2013 Aaron Adair

Published by *Onus Books*

Printed by Lightning Source International

All rights reserved. No part of this publication may be reproduced, stored in a retrieval system, or transmitted in any form by any means, electronic, mechanical, photocopy, recording, or otherwise, without the prior permission of the publisher, except as provided for by UK copyright law.

Cover image, "Under the Milky Way" by Steve Jurvetson (2007), used under Creative Commons licensing.

Cover design: Onus Books

Revised Standard Version of the Bible, copyright 1952 [2nd edition, 1971] by the Division of Christian Education of the National Council of the Churches of Christ in the United States of America. Used by permission. All rights reserved.

Trade paperback ISBN: 978-0-9566948-6-7

OB 05/09

PRAISE FOR *THE STAR OF BETHLEHEM: A SKEPTICAL VIEW*:

"Aaron Adair's investigation of The Star of Bethlehem doesn't just debunk a beloved Christian legend – it shows what a rough and treacherous neighborhood is that problematic intersection where scripture and science meet. Extensively researched, Adair's keen detective work uncovers a treasure-trove of both actual astronomical history and wild and wooly folklore; including the intrigues of Han Dynasty astrologers, the Antikythera Device - an ancient Greek clockwork computer, Hipparchus of Rhodes' discovery that the entire universe was shifting on its axis (or was it?), imaginary worlds like Immanuel Velikovsky's mysterious Planet X and pseudo-scholar Zecharia Sitchin's pseudo-Sumerian planet Nibiru. A fascinating and readable feat of hardcore historical legwork and keen scientific analysis."
—David Fitzgerald, author of *The Complete Heretic's Guide to Western Religion: The Mormons*.

"Well researched, scientifically reasoned, elegantly concise, this book will long be required reading on the 'Star of Bethlehem'. Full of fascinating historical facts, and better informed and more careful than any other book on the subject, this should be on the shelf of everyone interested in that legendary celestial event."
—Richard Carrier, Ph.D., author of *Proving History: Bayes's Theorem and the Quest for the Historical Jesus*.

"*The Star of Bethlehem* is a tightly-argued, well-reasoned exploration of a fascinating question. The explanations for the Star of Bethlehem are as varied as the people who proposed them, and Aaron Adair masterfully demonstrates why every effort to rationalize the Star thus far has failed—and offers the best explanation yet for what the Star really meant to the earliest Christian writers. *The Star of Bethlehem* is a concise and rigorous must-read for anyone interested in religion, history, and modern efforts to understand the past."

—Jason Colavito, author of *The Cult of Alien Gods*.

"Aaron Adair's book *The Star of Bethlehem: A Skeptical View* offers readers a fresh and interesting look into an old academic question. While the argument that the 'Star of Bethlehem' story is a myth isn't a new one, Aaron Adair—an astronomer and physicist at The Ohio State University—offers a look into the past through the eyes of a scientist, while not once ignoring the value of New Testament scholarship. This is a must-read, and perhaps the definitive, book on this subject."

—Thomas Verenna, co-editor of *'Is This Not the Carpenter?' The Question of the Historicity of the Figure of Jesus* and undergraduate student at Rutgers University.

*To Sarah, 'mein Schatz',
for being with me
when both near
and far away.*

About the author:

Aaron Adair is a recent PhD in physics from the Ohio State University, primarily working in the area of physics education research (PER). Previously, he studied physics, astronomy, and mathematics at Michigan State University. His scientific interests have included working as a show presenter at a planetarium, researching at the SETI Institute, and examining parts for the ATLAS particle detector at CERN. His current areas of research are technological innovations in the classroom, project-based learning curricula, the origins of physics misconceptions, the history of astronomy, and classical history and religions. He has published on the Star of Bethlehem previously in *Sky & Telescope* and *Zygon: Journal of Science & Religion*. Born and raised in Michigan, Aaron currently resides in Columbus, Ohio, but he is on the move, to where no one knows.

He blogs at gilgamesh42.wordpress.com.

Acknowledgements:

This volume has been in the making for years, but its creation was helped by many different friends, family, and teachers in many fields. I cannot thank all of them, but some deserve particular attention. The professor that more than anyone else brought me into the classical world was Debra Nails and her classes of Plato, Aristotle, and classical Athens. In order to engage in the texts, I needed to learn the Greek language, in which Carl Anderson and William Blake Tyrell have helped me, though I dare not claim proficiency as they can. In history, Linda Johnson showed me many of the methods of investigation, but a large debt also goes to Richard Carrier when it comes to studying ancient Christianity and with particular help related to this project. Robert M. Price has also been of great help, including publishing a paper on this subject in his (now defunct) journal. When I worked at a planetarium as a show presenter, which is also where I learned of these Star theories, I received help and support from my co-workers Don Batch, Shane Horvatin, John French, and Mary Gowans. When it came to these matters, I also had several times received direct help from David Hughes, who has one of the better books on the subject of the Star of Bethlehem, though his conclusions are very different than mine. My first publication on the Star was encouraged by my astronomy senior thesis advisor, Mark Voit, who also taught me much about modern research tools in cosmology. David Fitzgerald has been of great support in this project, and he got me in contact with the person that became my editor, Jonathan MS Pearce. It is because of his efforts that this volume is orders of magnitude better than it would have otherwise been. Lastly, I must thank Sarah Heim, who has had to hear the most about my research and has given me what I needed to hear in return and so much more.

Contents

Foreword 1

Preface 3

PART I: THE STORY 11

Chapter 1: Matthew's Account and Timelines 11

Chapter 2: Methodological Difficulties 19

PART II: THE HYPOTHESES 31

Chapter 3: Comets and Meteors 31

Chapter 4: Novae and Supernova 43

Chapter 5: The Planets and their Positions 51
    Part A: Conjunctions 53
    Part B: Horoscopes 69

Chapter 6: What (Other) Stars do Spangle Heaven with Such Error? Last Ditch Efforts 83

PART III: FATAL PROBLEMS 91

Chapter 7: Failure of all Natural Hypotheses 91

Chapter 8: Historical Issues 107

APPENDIX: THE TEXT OF MATTHEW 129

GLOSSARY 135

BIBLIOGRAPHY 147

Foreword

Like many astronomers, I've always been intrigued by the Star of Bethlehem.

It's the best-known and most legendary star, bar none. People who cannot name a single luminary in the actual night sky readily acknowledge their familiarity with the Bethlehem Star, and its religious role. No wonder it has inspired vast scientific speculation. For more than a century, various natural explanations have gained and then lost fashion, periodically making headlines. Moreover, it has been a staple of holiday planetarium shows since the 1930s.

For me, personally, this celestial topic launched a quarter-century career. My "Night Watchman" and then "Strange Universe" pages have appeared in every issue of first *Discover*, and then *Astronomy* magazines. But the very first column, published in *Discover* in December 1989, was a two-page spread about the Star of Bethlehem.

Basically I summarized the various "explanations" shown to the public during planetariums' annual "Star of Wonder" shows, then noted that Planetarium Directors—I'd interviewed quite a few—were well aware that each was impossible. Nonetheless, the shows remain popular, and have become such a tradition in and of themselves that no one seems bothered by such make-believe science being annually offered to the public.

I did not have the research time, nor found it appropriate, to delve into various other aspects of The Star. How, for example, the nativity account in Matthew is so utterly different from the gospel of Luke. How astrological explanations might (or might not) solve the puzzle. Or the hard evidence behind the idea that Matthew—a century after the fact—may have simply created the Star story by borrowing specific aspects from popular contemporary literature.

Aaron Adair does all this and much more, and does so in an immensely readable way. The result is not a rehash of previous research into this timeless subject. Rather, it is a wonderfully meticulous, thoroughly engaging "final word" that leaves the reader not with nagging questions or dangling what-ifs, but a strong sense that the case may finally be closed.

Dr. Adair dissects each science "explanation" with astronomical insights both accurate and inarguable. Refreshingly, he does so without in any way disparaging religion in general or Christianity in particular. He does not cross the hubris line as a few popular physicists do, by suggesting that as a scientist, he possesses a superior "take" on theology. Nowhere do we feel smugness, or regard his brushstrokes as too wide.

For anyone truly interested in that most famous Star in history, this book is essential reading.

Bob Berman
Willow, New York

Preface

"Finally, the clouds are parting. I may get to do some observations tonight!" he thought while shivering, his face cold save his beard which had reached new lengths. He had been monitoring the skies quite avidly for the last year or so. Perhaps he was so keen to get out and look, taking every opportunity he could find, because the skies were particularly poor for star watching where he was. But there were others like him who believed that the stars had been boding for something quite remarkable. In the previous year, there was a rather ominous conjunction of the planets Jupiter and Saturn, who in myth were not friendly, though they shared Titanic blood. Could this mean there was new blood to be spilled? Was there to be another war with the king's neighbors? Or even a civil war?

The signs were becoming even starker as Jupiter and Saturn were now massing with a third wandering star, forming a triangle in the night sky, the likes of which had not been seen for some time. The only object, the Arabs thought, that could make the great conjunction and massing even more potent a portent was the sight of a comet. Astrologers about the land were most curious if something they could not predict with their charts was to happen.

And so this one astrologer was out in the chilly October night, perhaps finding a sign of things to come. He knew what to look for, or so he believed. After all, he had mastered and taught his craft for years, though his genius was not recognized by his bored students. But political figures respected him and sought his judgments, which the astrologer was reluctant and nervous to give. Did he really trust his art? Was there any wheat in all so much chaff? Diligently he worked out new ways to advance his field, saving it from ridicule the best he could. If anyone could save this art, many believed it was him. He was recognized for his scientific prowess, and his mathematical abilities were first-rate. In recent years, he pored over data from ages past, trying to find patterns, and occasionally an abrupt "Eureka!" would stir him away from the rest of the world. His overwhelming insight from nine years earlier about the orbits of the planets still pushed him, and tonight was one of those nights that he would look into the heavens and be lost to all else.

As the clouds broke, the familiar stars were beginning to be seen. It was still too early to see the brightest star of all, Sirius, while the constellations of spring were well past their zenith. Winter was coming, and with it many another constellation that told of coming storms and conflicts. The north was the only constant, at peace, and yet there was the terrible dragon forever chasing about the North Star all day and night. Only when God finally ended it all would such evils be replaced in the new heavens and new earth. Or so he pondered as infinities floated about his mind. Along with that circle in the north, the astrologer could see the constellation that would point to Polaris, a collection of stars which some called a bear, and if he could just squint right he could see the double-star down its back and tail. A rider on a horse, some say, forever circling the north. Horses on bear tails; dragons spinning about; a thousand fantasies all trapped in the unchanging firmament of the north.

"Enough of this," he blurted to himself. What would those like Caspar think of him if he lost himself there?

He was not supposed to focus long on what was constant, as it was change he was interested in seeing, including the change in the wanderers, the planets. Perhaps even more had changed, if the rumors were true.

The gaps in the clouds had expanded to the south, the zodiac undressing before his eyes, a body of stars he had come to know intimately over the years. At first, to the east, the beautiful Andromeda was uncovered. When he squinted he could still see a bit of unmoving cloud at her hip. Curious, but it was not the place to find any of the planets. Tonight he needed to concentrate elsewhere. With the skies clearing, he could see the constellation of the serpent-bearer, what the Greeks called Ophiuchus. The Greeks also thought him a healer, an Asclepius. Was this pattern of stars telling us of the resurrection of the dead then, or was the serpent to poison the land? Tonight the snake handler would also be bearing planets and prophecies.

And there he saw it: amongst the massed planets was something new and bright. Between Saturn and Jupiter, who had just returned to the Fiery Trine of the astrologers, there emerged a new flame, a new star. Had the prophecy of the Arabs come true? The astrologer carefully wrote down his observations to later share with his peers what he deemed may come about. A great tumult was had amongst them because of the newness of the star. Some prophesied doom, others revolution, others still the End of Days. While religiously important to him, the key astronomer of his country was not so quick to interpret.

The cautiousness of the experienced scientist was not simply a quirk of personality, but he had read of other stories of new stars and what they could have meant. He was a dabbler in books of history and religion, understanding both present and past with his craft. In his consumption and obsession, he consulted rare books concerning the powers that ruled a small territory on the Mediterranean, the land of Judea.

A new star. The lands of the Jews. Ancient prophecy. Writing to his benefactor, he began to pen his interpretation of the brilliant phenomenon. Perhaps at last he understood the truth...

This story may seem an interesting tale that would connect to one recognized by literally billions of people today. As many know, in December (and January for some) there is the celebration of Christmas, the time of the birth of one of the best-known religious figures, Jesus of Nazareth, called Lord Jesus Christ by the reverent. Part of the story of his birth includes the coming of men from the East who had seen some star in the sky that told them of the birth of the King of the Jews. And many have wondered what these men, called Magi (sometimes translated 'wise men'), could have seen that would have inspired them to travel from their native land to the hamlet of Bethlehem in order to see and pay homage to an infant.

But the story above is actually not about the Magi. It is a fictional retelling of the work of a more modern and well-known figure, Johannes Kepler. This German astronomer not only revolutionized the sciences of his days, best known for the discovery of three laws of planetary motion, he was also interested in what is now been soundly debunked within the scientific community, astrological influences. The observations in question concern the conjunction of Jupiter and Saturn in 1603 and his eponymous supernova of 1604. It was also around this time that Kepler came across scholarly arguments that would put the birth of Jesus not in 1 BCE per tradition but 4 BCE or earlier. Kepler also looked into astronomical circumstances at the new time of Jesus's birth, and one association amongst astrologers was the conjunction of, again, Jupiter and Saturn, known to have taken place in 7 BCE (based on calculations).[1] This part of the story will be returned to later on.

The efforts of Kepler have been used by many a modern researcher to amplify their own search for what was of such interest to the Magi now over 2000 years ago. Called the Christmas Star or the Star of Bethlehem, this object has garnered an immense amount of speculation about what astrological readings could have been of interest to the eastern 'wise men' and what the Star itself was.[2] But even though Kepler is seen as the light that many Star researchers follow, the founding figure of modern science believed the great luminary at Jesus's birth to be a miraculous object, not some distant planet, star, comet, or anything of the sort.

But Kepler believed in astrology and other miraculous things, so he was wrong about plenty. Perhaps the Star of Bethlehem, as described in the Gospel of Matthew, can be explained by some natural phenomenon that can be uncovered using modern astronomical erudition. However, in this age of greater scientific knowledge, there is also an increase in historical knowledge and methodology, and the latter subject cannot be overlooked, though it often is. Should we not

---

[1] The most helpful biography of Kepler remains Max Caspar, *Kepler*.
[2] For example, Ruth Freitag, *The Star of Bethlehem: A List of References* is over 20 pages of, as the title suggests, just references to articles and books with little or no summary, published in 1979; since then, the book could be nearly doubled in length.

ask "Was there a Star at all?" before we try to explain it? Ancient books tell of all sorts of things we are, at the very least, skeptical of, and the Bible has not been free of such inquisitions by skeptic and Christian alike.

The Star of Bethlehem is a subject that touches two of the chief intellectual forces in the world today: science and religion. Arguments to their compatibility have lasted for many decades, and part of that discussion has included events such as the Star. If it could be verified by science, it would show a dialogue between the paradigms of thought rather than conflict or avoidance. Exploring this debate can potentially reveal how the ideas of faith fit into the realm of empirical inquiry, not to mention shed light on history and one of the most influential figures in modern times.

This volume will focus on the efforts trying to explain what the Star was and why someone in antiquity would have thought it so special as to signify the birth of a king in Judea. This book can be considered a companion volume to Jonathan Pearce's *The Nativity: A Critical Examination*, which considers matters of historicity for both versions of the canonical birth of Jesus stories more than can be done here. There, the Star was considered for several pages, but this book will be more thorough in showing the various theories put forward, and their difficulties (something that I had done in the past[3]). Moreover, the current book is a summary of a larger planned volume about the Star along with the historicity of the Nativity in the Gospels and the literary context of the story.

To approach this properly, the book is organized in the following way. First will be a summary of what the

---

[3] Aaron Adair, "Science, Scholarship, and Bethlehem's Starry Night", *Sky & Telescope* 114, 6 (Dec 2007): 26-9.

second chapter of the Gospel of Matthew has to say about the Star. From that starting point some basics of the chronology will be considered so we know when in time anything of interest ought to be. After this, in Chapter 2 some of the methodological problems with any attempt to find the Star of Bethlehem will be discussed, including what sort of source Matthew is. From the particular details to the historical problems at hand, this chapter will show that the effort is rather fruitless even assuming the historicity of the account. From there, each following chapter will discuss a particular hypothesis that has been put forward over the decades of what may fit the description of Matthew's account; however, Chapter 6 will discuss a number of other, less popular hypotheses but have been published and discussed in the Star of Bethlehem literature nonetheless. In those chapters about the various Star hypotheses, the major problems each theory has will be demonstrated. The penultimate chapter will demonstrate that no natural phenomenon can fit Matthew's account, leaving it up to the miraculous (or perhaps nefarious ETs and their ships). Finally, several points will be levied against the historicity of this particular part of the story of Jesus's birth. Even if the miraculous is allowed and it does happen, the last chapter should demonstrate that this account is just not credible.

A few more notes should be made here. First, when referring to the stellar object from Matthew's story, the word 'star' is capitalized. Similarly this is done for canonical gospels, the version of the nativity from the Bible, and other titles associated with the Christian canon. As for the text of the Gospel, I tend to use to Revised Standard Version (RSV) as a guide, though in many cases I use multiple translations and consult the original language directly. My own translation of the story can be found in the appendix of this volume. For dating, I have tried to be consistent with scholars that use the convention of Before Common Era (BCE)/Common Era (CE), which are equivalent to the better known BC/AD system.

When it comes to terminology, I will explain technical terms as they appear, and a glossary of various terms is provided at the end of the book. However, it is worth distinguishing here what is meant by 'astronomy' and 'astrology'. While separate terms were not used in antiquity, they have meanings that are rather important for modern scientists and historians. Astronomy is the study of the stars and planets, including their motions. Astrology is a form of divination using the positions of the stars and planets. In the past, the astronomers and astrologers were the same people and profession, while today one can be an astrologer with minimal knowledge of celestial mechanics. A modern astronomer, on the other hand, is unlikely to even talk about astrology, let alone practice it. (If you chat with someone and they tell you they are an astronomer, one sure way to anger them is to ask what their Sun sign is.)

With these considerations in place, let us consider what has been one fascinating intersection with science and the Christian religion.

PART I: THE STORY

Chapter 1: Matthew's Account and Timelines

The oldest version of the story we have about the Star of Bethlehem comes from the author of the Gospel of Matthew, the first book in the New Testament of the Holy Bible. While carrying the name of one of the Disciples of Jesus of Nazareth, scholars have been in agreement that no one really knows who wrote this book. There is also considerable uncertainty as to where and when it was written. Many introductions to the New Testament and similar resources[4] will mention that it may have been written in Antioch on the Orontes River, then in the Roman province of Syria (its ruins today rest in Turkey), but this cannot go beyond a reasonable speculation.

The dating is also said to be around 85 CE, which would already place the writing of the book nearly a century after Jesus's birth. However, the date is uncertain. Generally it is agreed that Matthew uses another book as its source, the Gospel of Mark, and that

---

[4] i.e. John Barton and John Muddimen, *The Oxford Bible Commentary*, pp. 844-845; Raymond Brown, *The Birth of the Messiah*, pp. 45-48.

was written probably sometime after the Jewish Revolt from 66 to 73 CE. Mark (or whoever wrote the Gospel) mentions the destruction of the Second Temple in chapter 13, and that structure was destroyed by the Romans in 70 CE. This marks, then, the *terminus a quo*, or earliest date, for Mark, and hence the earliest for Matthew who used Mark. The author of Matthew also mentioned the Temple's destruction, so this independently forces a post-70 CE dating. As for the latest possible date, the *terminus ad quem*, this has more contentious scholarship surrounding it. For simplicity, I will point to the most unambiguous early citation of Matthew's Gospel, Justin Martyr, who wrote his works around 155 CE. He refers to the Gospels as the "memoirs of the apostles", but his citations show that he is talking about the texts we know of. He also explicates the Star of Bethlehem in his *Dialog with Trypho* (chapters 106 and 126), connecting it with various parts of the Hebrew Bible. This means that, broadly, Matthew was written between 70 and 155 CE.[5]

At this point, it is worth mentioning the writings of Ignatius of Antioch, a bishop in the second century. His letters, alleged to have been written while a prisoner about to be executed, have been pointed to as earlier citations of Matthew, but this is quite problematic. There is uncertainty that the letters are authentic, that Ignatius lived and wrote when he is said to have done (his martyrdom is said to be during the reign of Emperor Trajan), and if he actually cited any Gospel. None of

---

[5] David Sim, "Reconstructing the Social and Religious Milieu of Matthew: Methods, Sources, and Possible Results", in *Matthew, James, and Didache*, eds. Huub van de Sandt, Jürgen Zangenburg (Atlanta: Society of Biblical Literature, 2008), pp. 15-19 notes that the conflict with formative Judaism seen most clearly in Matthew would make better sense in the second century.

these things can be decisively shown, and there is considerable scholarship contesting all of these points.[6] The arguments concerning authorship seemed to have the greatest consensus (that they are authentic), but the rest is arguable. Other alleged limits from sources such as the early second century bishop Papias and the Christian treatise called the *Didache* are similarly problematic. Allegedly Papias said Matthew was a collection of sayings in Hebrew,[7] but our Gospel is in Greek and far more than a list of sayings, while the *Didache* also fails to say it quotes from a Gospel, let alone Matthew. As such, there is minimal utility in looking to these sources to shrink the time that Matthew wrote.

For now, these details matter little. The fact that Matthew wrote in 75 or 175 CE will not be a significant factor with regard to the conclusions of the following chapters. At any rate, Matthew is writing generations after the birth of Jesus (assuming his chronology), and what matters are his methods and sources. After all, centuries after the fact we can write decent biographies of Benjamin Franklin because of the sources at hand, including his own writings. If Matthew had good sources and competently and critically used them, it matters not one iota when he wrote. On the other hand, if his sources were poor and his methods unsound, it wouldn't matter if he wrote at the earliest possible date.

---

[6] Timothy Barnes, "The Date of Ignatius", *The Expository Times* 120, 3 (2008): 119-130 notes recent arguments for the inauthenticity of the letters, the incorrect dating of the letters, and his own argument that they fit best in the 140s or so. On the independence of Ignatius from Matthew, see William R. Schoedel, *Ignatius of Antioch*, p. 9.

[7] Eusebius, *Ecclesiastical History* 3.39.16.

However, for the time being we will not question Matthew's account or bring up any inconsistencies it has with other sources about Jesus, something that Jonathan Pearce does in his *The Nativity* as well as Raymond Brown in his *Birth of the Messiah*. Let us for now consider what Matthew says.

The story of Jesus's birth is begun by the Evangelist in a rather boring way with a list of who-begat-whom, but then he gets into the narrative proper, starting with the unusual circumstances of Jesus's birth. Then, in the second chapter, we come to the story of the Magi. During the reign of King Herod the Great (37-5/4BCE), these unnumbered eastern sages came to Jerusalem saying that they had seen the star of the King of the Jews rising. Some translations say the Star was "in the east", including the King James Version, but more correctly the Greek refers to the rising of the Star, which would be on the eastern horizon, and this is reflected in several modern renditions into English. Now, the Magi believing the king to have been born, they wanted to worship him. Herod, who was the king and had been for some time, was justifiably troubled by this. He further inquired into the Star the Magi talked of, even calling on the religious leaders in Jerusalem to help figure out where the Messiah (apparently it was Herod that connected the king and the messiah together; these titles are not synonymous) was to be born. According to prophecy, he was to be born in Bethlehem, just a few miles south of Jerusalem. Herod then told the Magi to go find the Messiah and inform him of his whereabouts so that he, too, could worship him.

The Magi left, and then the Star appeared to them (perhaps it had disappeared?) and "went before them until it came and stood over where the child was" (Matt

2:9). Here is the story at its most spectacular and which all natural Star hypotheses must contend with. The Star "goes before" the Magi, arrives, and then stands over a particular locale. At this point the Magi are at Bethlehem, perhaps just outside the very house Jesus and his family were in (the Gospel specifies it was a house that the Magi enter), and they then enter in. They see the child with his mother, worship Jesus, and give gifts of gold, frankincense, and myrrh, all items of considerable value. From there, the Magi stay awhile and then have an angel telling them to leave town; Joseph has a similar message in a dream, and soon all the good guys in the story have escaped from Bethlehem and the oncoming wrath of Herod.

So, the sorts of details that the Star has, according to Matthew, are as follows:

- There was a star or something star-like (2:2, 9)
- It was visible to the Magi ("for we have *seen* his Star at rising" 2:2)
- Found in the east/at rising (2:2, 9)
- Told of the birth of a king/messiah (2:2, 4, 11)
- Went "before them" until... (2:9)
- Stands over "where the child was" (2:9)

But there remains one last detail that isn't explicit, and that is the date range for the Star, at least according to Matthew. What we are told is that Jesus was born before the death of King Herod. The details of the story also imply that the Holy Family stayed in Egypt for a relatively short time, after which they returned to the Holy Land when Herod died in 5/4 BCE. (Some have argued that Herod died later, some even claiming as late as 1 CE, but I will remain with the standard chronology as it fits the evidence best.) But this does not strongly establish by what date Jesus would need to have been born by.

At this point many scholars will turn to the Gospel of Luke which gives further details (though he says nothing of stars or magi at the Nativity). For example, in Luke 3 we are told that John the Baptist began to minister in the 15th year of Emperor Tiberius, which would probably be 28/29 CE. Jesus came to John sometime after he began to baptize by the River Jordan, and Jesus was "about thirty" (Luke 3:23) when he came to John. Taking 28/29 CE as the earliest date that Jesus came to John and assuming Jesus was 35 years old, then that would put Jesus's birth in 8/7 BCE, the furthest we can go back given the Lukan information. But this assumes that Jesus started his ministry in the same year John began to preach, and Jesus was as old as possible given the language of Luke. We are at least as justified in thinking Jesus started his ministry in 34 CE at the age of 28 and thus born in 6 CE. Moreover, the above strained chronological speculation ignores what Luke does say about when Jesus was born, which was during the census of Quirinius when he was governor of Syria (Luke 2:2) which was in 6/7 CE. This is a problem discussed elsewhere, so it will not be argued about here; we only need to note the difficulty in establishing any early biographical information about Jesus and how we cannot simply blend our sources together indiscriminately.

Nonetheless, the context of Matthew is the best guide we have when it comes to dating the Star, and it implies Jesus was very young when his family escaped from Judea to Egypt and returned (though to Galilee) after a short time. So, from that we could leave a broad period of time consistent with Matthew's account between 10 and early 4 BCE for the birth of Jesus. But this is merely a best guess, and it ignores other information we have. Because of this difficulty, a date earlier than 10 BCE is not necessarily out of bounds, though we should remain strict about the 4 BCE date as we are more solid about when Herod died and Jesus needing to be born before that, according to Matthew.

Chapter 2: Methodological Difficulties

Before we investigate what natural event may have been behind the Star of Bethlehem, there are some methodological problems that need to be highlighted which are not often well-considered in the literature. Part of that list of problems was discussed in the prior chapter, but they will be highlighted again here.

Firstly, we are rather unsure about the dating of Jesus's birth. None of the canonical Gospels provide an exact date, nor even a year, so we are forced to infer what we can from those books. The historical context Matthew provides in his story can only provide a *terminus a quo* with the death of Herod. The data from Luke could help establish a *terminus ad quem* based on the data when Jesus came to John, as noted in the previous chapter. However, that must ignore the information about Jesus being born during the 6/7 CE census of Judea when it came part of the province of Syria under Quirinius. In the Star of Bethlehem literature, the way this is dealt with is varied. For example, astronomer and historian David Hughes,[8] in his book, looked at some of the attempts to reconcile the accounts of Matthew and Luke. Retired astronomer Michael Molnar,[9] on the other hand, is not convinced by these attempts and sees it as a serious issue, though this does not seem to make him skeptical of the account

---

[8] David Hughes, *The Star of Bethlehem: An Astronomer's Confirmation*.
[9] Michael Molnar, *The Star of Bethlehem: The Legacy of the Magi*.

of the Star in Matthew. However, without either a reconciliation or a demonstration of which dating of Jesus's birth is more reliable (assuming either are reliable), then searching for an astronomical event or astrological circumstance is pointless unless something very definite can be cited as to what gave rise to the story of the Christmas Star.

The chronology is also confusing to modern researchers because Luke's version of the birth of Jesus has very little in common with Matthew's account. Most strikingly, Luke has no mention of the Magi, the slaughter of babies in Bethlehem, the flight to Egypt, or the Star. Luke is also the only Gospel author that even puts on the mantel of a historian, so the absence of the amazing story seen in the First Evangelist should be alarming. Did Luke not include the story because he thought it wasn't historical? Or did he just not hear of the tale? And if he didn't know of the story, and Luke claims to be using witnesses of the Jesus tradition (Luke 1:2), what does that mean about the reliability of Matthew's tale of astrologers following a strange luminary?[10]

This leads to the second methodological problem, one perhaps even more insurmountable than the prior one. The premise of all Star scholarship is that something in the skies indicated to the Magi that a king was born, specifically Jewish. But this requires delving into astrological interpretation, and that is extremely problematic. This is not because astrology is a failed

---

[10] The debate about the relationship between Matthew and Luke is long and nuanced, but the arguments by Mark Goodacre in *The Case Against Q: Studies in Markan Priority and the Synoptic Problem* are suggestive that Luke did know and use Matthew for writing his Gospel. However, this possibility will not be considered here, and I reserve a greater discussion on this for future work.

science (which it is), but because portents can be almost infinitely interpreted to mean most anything. This has been noted in scientific studies of the subject as practiced by modern astrologers.

For example, consider the following study with several astrologers, all of whom were respected by their peers and had practiced their craft for years. They had participants answer questions about personality traits that fit what the astrologers wanted for analyzing their horoscopes, along with the date, time, and location of birth of the participants. The results were then mixed up and the astrologers had to match up the 23 participants' horoscopes with their personal details. The study found that the astrologers could not connect the horoscopes with the personality reports any better than chance, as expected of a subject that does not have any basis in reality. However, more interesting is the fact that the astrologers were in amazing contradiction among each other when it came to which horoscope went with each personality report. That is, one astrologer would say horoscope A fit with personality report B, but another would say horoscope A fit personality report C, and so on with the remaining personality reports. On average, the agreement between any two astrologers was only 1.4 times out of 23. That is basically at the level of chance.[11] A Dutch study of similar design found much the same results: accuracy was about at the level of chance and so was agreement between astrologers.[12] So, at least for modern astrologers, what a given horoscope will mean is

---

[11] John McGrew and Richard McFall, "A Scientific Inquiry into the Validity of Astrology", *Journal of Scientific Exploration* 4, 1 (1990): 75-83.
[12] Rob Nanninga, "The Astrotest: A Tough Match for Astrologers," *Correlation* 15, 2 (1996): 14-20.

just as determinate as a roulette wheel. This makes any interpretation of past skies highly suspect since most any condition of the sky could be said to mean almost anything.

This condition can be seen in the ancient records as well, including those from Christians. For example, Christian critics of astrology said that the astrologer simply looked for what their client wanted to hear as any horoscope had positive and negative conditions; just pick the parts you wanted.[13] Pagans such as the historian Tacitus were of a similar opinion.[14] In his critique of astrology, the famous Roman politician Cicero gives the example of a Babylonian astrologer who made predictions for the three members of the Triumvirate—Crassus, Pompey, and Caesar. The astrologer, apparently wanting a happy story for these figures as well as the Roman public that depended on their alliance, stated that all three would live long lives and be glorious. Cicero wrote after the assassination of Caesar, which itself followed the failed invasion of Parthia by Crassus and his demise as well as the civil war between Pompey and Caesar that ended with Pompey's death.[15] For Cicero, it was obvious how astrologers were poor prophets and instead brownnosers.

Astrologers of the era were well aware of the problem of poor predictions. The most famous of the ancient Western astrologers, Claudius Ptolemy, wrote in the second century that the subject had many fraudsters, but even the best will make mistakes because of the

---

[13] Cf. *The Recognitions of Clement* 10.11; Origen, *Philocalia* 23.21-2.
[14] Tacitus, *Histories* 1.1.22.
[15] Cicero, *De Divinatione* 2.99-100.

immense complexity of the task,[16] an argument repeated by a fourth century Christian convert and astrologer, Julius Firmicus Maternus.[17] In other words, because there are so many things that can be considered in putting together and interpreting a horoscope, experts are bound to disagree.

So, the ancient astrologers in the Western tradition already tell us that the act of interpretation is very difficult, and the evidence is that astrologers cannot agree even when they all follow the same rulebook. The work of the historian of astrology Tamsyn Barton also shows that the ancient treatises of ancient Western astrologers (such as from Marcus Manilius, Ptolemy and others) actually do not tell us how to interpret a star chart.[18] But more damning is the incredible level at which astrologers will contradict each other and even themselves. One particular example should be examined because of its importance to this study: what constellation or astrological sign corresponded to or influenced what region of the world; this is known as geographical astrology. This is important for Star research because of efforts to find the constellation which would have been associated with Judea. The only specific list that mentioned Judea comes from Ptolemy who connected that region with the zodiac constellation of Aries. However, a century earlier in the writing of Manilius this same general region is under the purview of Aquarius. Later in the same century, Dorotheus of Sidon goes with Gemini.[19] There are still other astrological geographies,

---

[16] Ptolemy, *Tetrabiblos* 1.2 (6-7).
[17] Firmicus, *Mathesis* 1.3.2.
[18] Tamsyn Barton, *Ancient Astrology*, pp. 114-142.
[19] Cf. Wilhem Gundel and Siegfried Schott, *Dekane und Dekansternbilder: Ein Beitrag zur Geschichte der Sternbilder der Kulturvölker*, p. 312.

and none agree more than once or twice with each other and none agree on any one thing.

As such, historians of astronomy and astrology, such as John North (who studied ancient and medieval horoscopes), would not try to interpret a given horoscope since how it was done was subject-dependent; there are too many ways to read a star chart.[20] Considering how much contradiction there is between astrologers on so much, from methods to meaning, there is no respectable position one can take on how a horoscope would have been interpreted in antiquity. But this problem is compounded even more by another enigma: which version of astrology is being used?

There is not one, singular version of this divination technique, and there are significant differences between how astrology is done in the Western tradition compared to the Chinese version. However, even the Western tradition is taking on an earlier form of the art from Assyrian and Babylonian traditions which exist in records of great antiquity. We have some horoscopes from these lands and times, but what has been seen is that it is not the same as the methods better known from Greek and Latin treatises from the second century BCE and onward. One of the defining features of the Western horoscope is the arrangement of the cardinal points. These are the locations of the eastern and western horizon (the ascendant and descendant), along with the point on the zodiac that is highest in the sky at the observer's location and the point 180 degrees opposite on the zodiac (the midheaven and anti-midheaven). The location of these points in a horoscope are very important to interpretation, and considerable calculations are

---

[20] John David North, *Horoscopes and History*, p. xi.

done to find and place them, yet they are absent from the older Babylonian versions of horoscopes. Moreover, we have no good information to help us even guess how ancient, eastern horoscopes were interpreted as most of them failed to include an interpretation, and we have no treatises on the subject.[21] The closest is the astrological text called the *Enuma Anu Enlil*, written sometime around 1000 BCE, but how this source was applied in the first century is almost impossible to extrapolate.

This means we have almost no basis to discuss how any configuration of the stars would have been interpreted by eastern diviners. We also have to be more specific in which group one considers the magi. These pious figures were at first perhaps a tribe in modern-day Iran (the Medes), but by the rise of the Persian Empire it had become a priestly cast in the Zoroastrian religion. More discussion of the magi and their religion will take place in Chapter 8, but for now it is worth mentioning that the attitude towards astrology at the time of Jesus's birth is even more divergent than what we know from the Greeks and Babylonians. Overall, the hurdle of interpretation of astrological portents by moderns ought to be out of the purview of appropriate method.

Lastly, we should consider an issue that is poorly addressed in the literature by those searching for the Christmas Star: what makes us think the author is trying to tell us history and why ought we trust him or his sources (if he had any concerning the Nativity)? Besides having the correct reconstruction of the text (something which we are in reasonably good standing when it comes to Matt 2; see the translation provided in this book), we need to consider the literary purpose and

---

[21] Francesca Rochberg, *Babylonian Horoscopes*. Transactions of the American Philosophical Society vol. 88, pt. 1.

context of a text. Are we dealing with a diary of events similar to Julius Caesar's *Gallic Wars*, a biography based on eyewitnesses such as Arrian's history of Alexander the Great (the *Anabasis*), or do we have folklore and romances as in the stories of King Arthur? When it comes to the Gospels, this is not an easy question and has created mounds of collected paper and ink, all with both helpful and unhelpful observations, insights, and points of method in determining what we are dealing with.[22] But perhaps key is the fact that the story of the amazing portent at birth indicating greatness of the infant is part and parcel of stories told of heroes, kings, and gods in antiquity. This includes stellar features as well. Famous stars leading heroes to their destination include that of Timoleon on his way to Sicily[23] and Aeneas on his way to Italy.[24] We also have the parallels between Jesus's birth and Moses's, especially in non-canonical sources concerning the latter (which include a star at his birth interpreted by 'wise men' of sorts).[25] Put simply, how does one categorize the story? You cannot skip this step and do proper analysis; else you will simply misunderstand what you are reading.

But even assuming, upon reflection, that Matthew's account is a sober one recorded by Joe Friday trying to get "just the facts", we need to figure out what his sources were, what quality they were, and how critically Matthew treated them. Unfortunately, unlike other

---

[22] On the genre of at least some of the Gospels, the most useful argument that has a theoretical foundation comes from biblical scholar Michael Vines in *The Problem of Markan Genre: The Gospel of Mark and the Jewish Novel*.

[23] Plutarch, *Timoleon* 8; Diodorus, *Biblotheca Historica* 16.66.3.

[24] Virgil, *Aeneid* 2.687-711. See Chapter 8 for more discussion of this tale.

[25] John Bowman, *Samaritan Documents: Relating to their History, Religion, and Life*, pp. 287-288.

historians and biographers of the time, including the lower-quality ones, Matthew does not tell us what his sources were, nor does he do anything like a critical examination of them. Often it has been a premise of form criticism, the study of the conveyance of stories and sayings found in the Gospels, that the stories about Jesus came through some sort of oral tradition from his early followers and then through a chain of transmission (be it via plebs or pastors) to the authors of the Gospels. Unfortunately, we have great difficulty figuring out what may have come from history and what may have been good storytelling. This problem is compounded immensely for the Nativity accounts because at this time Jesus would have been little-known, having not started his ministry and gathered a following yet. As such, none of the Disciples of Jesus would have been witness to his birth. This leaves Jesus's family, at best, to tell the tale. However, this is further limited by the apparent disappearance of Jesus's father, Joseph, from the Jesus story after the Nativity. Tradition has it that Joseph was dead by the time Jesus began preaching, leaving only Mary as witness. But even she disappears at the beginning of the Christian movement, last mentioned in the Book of Acts and otherwise existing in later legend (legends that include her own miraculous birth). Any oral tradition then about Jesus's birth (if it existed) would not be guaranteed by a living witness when Matthew wrote, and we have no evidence to show the story could be traced back to any of those witnesses. But most importantly, Matthew does not tell us what his sources were and why he thought they were useful.

Compare that to a biographer from around the same time who is usually considered a less than trustworthy writer, Suetonius. He wrote biographies of the Caesars, starting with Gaius Julius and ending with Domitian, and for his earlier works he had access to the official Roman archives. Consider his biography of Gaius Octavius, later known as Caesar Augustus. He tells us various stories about the birth of Octavius, many of which include omens that are reminiscent of portents at Jesus's birth. However, Suetonius distinguishes these tales from that which he finds in various documents that he names. Suetonius also considers the plausibility of the folktales that surround the emperor. The same thing can be seen in Suetonius's biography of Emperor Nero, often considered overly credulous. It is from him that we learn the story of Nero playing his lyre while Rome burned, confusing gossip with what really happened (Nero wasn't even in Rome according to the more reliable Tacitus).[26]

Matthew, on the other hand, does not discuss his sources, does not separate witnesses, and provides not a hint of skepticism or critical engagement with the stories. The only source that he does cite is the Old Testament when talking of prophecies about Jesus, but if this is indicative of his efforts then Matthew isn't even a careful historian. As is well-known, he has either misinterpreted or misquoted parts of the Hebrew Bible or has even made them up (the prophecy that the Messiah had to be from Nazareth is not found, as is, in the actual Hebrew or the ancient Greek translation commonly used around this time; the prophecy of the birth in Bethlehem is also a confabulation between verses of different Hebrew authors). Since Matthew's historiographical methods are vastly inferior to that of an author considered suspect, and his access to reliable information is

---

[26] Tacitus, *Annals* 15.39.

even more remote, we have to, *a fortiori*, consider Matthew even more suspect for historical facts, and this assumes Matthew is even trying to tell us facts.

But these problems of literary context and historical method are not well-argued in Star of Bethlehem literature. For example, David Hughes, back in his 1979 book, spoke of the story by Matthew as having "the ring of truth",[27] but this can only be the affect fallacy. Little else is said by proponents to show that the tale likely has a kernel of truth. What this means is that most who approach the subject looking for the Star as a natural event have assumed a lot about the text and its historicity. The lack of awareness of what problems the narrative has before analysis will not likely be fixed by ignoring them, and Star researchers should be both aware of and deal with them.

Still, let us bypass these issues for now, though they are significant, and instead give these investigations into the Star their best shot. If someone could show a good correspondence between Matthew's account and some particular stellar configuration or series of events, then that could affect the probability of Matthew telling us something true. But even if a solid connection between events in the sky and Matthew's tale can be uncovered, we may still need to be skeptical of the historicity of the story because of the uncertainties about Matthew's sources. Other possibilities are back calculations by someone later on who used a few equations to figure out what sorts of interesting sights were in the sky around

---

[27] Hughes, *The Star of Bethlehem: An Astronomer's Confirmation*, p. 198. In private conversation, Hughes has expressed more skepticism of the story.

when Jesus was born,[28] or a similar event at a later date inspired the story.[29] These possibilities also need to be considered before reaching the conclusion that we have helpful data about the birth of Jesus rather than theologizing of natal astronomical events well after the fact.

But before coming to these decisions we need to see what natural phenomena are alleged to fit the tale. That is the subject of the following chapters, to which I hope you will now turn.

---

[28] Cf. Lynn Thorndike, *History of Magic and Experimental Science*, Vol. 1, pp. 471-472.
[29] Cf. Rod M. Jenkins, "The Star of Bethlehem as the Comet of AD 66", *Journal of the British Astronomical Association* 114, 6 (June 1999): 336-343.

PART II: THE HYPOTHESES

Chapter 3: Comets and Meteors

Perhaps the oldest argument put forward for a non-miraculous version of the Star of Bethlehem comes from a German Protestant theologian, Christian Gottlieb Kühnöl, in 1807. Coming from a wave of scholars in reaction to higher criticism of the Bible, he and others were starting to reimagine what the Bible said about miracles, and Kühnöl in particular focused on what could explain the Star. Not only being one of the first in this area, the author is also one of the published few that argued the Star could be explained as a couple of meteors.[30] More recently, the late Sir Patrick Moore, the voice of popular astronomy in Great Britain, in one book[31] and in BBC television conversations about the Star with David Hughes and Mark Kidger, favored the meteor explanation for the Star back in 2001.[32]

What advantages does this hypothesis hold? The flash of a meteor, which is basically a rock flying through the atmosphere that came from outer space, can travel

---

[30] Christian Gottlieb Kühnöl, *Evangelium Matthaei*, p. 23.
[31] Patrick Moore, *The Star of Bethlehem*.
[32] Martin Mobberley, *It Came from Outer Space Wearing an RAF Blazer!: A Fan's Biography of Sir Patrick Moore*, p. 535.

in most any direction, so it could have first been seen overhead in Jerusalem by the Magi, and that streak of light could have traveled south towards Bethlehem. So perhaps this can fit the description of the Star "going before" the Magi. Moreover, any year for the birth of Jesus can be consistent with a meteor hypothesis since they happen most all the time, and showers of them regularly happen throughout the year.

However, this is pretty much all the criteria a meteor could potentially fit, while it significantly fails to fit other details. For one, it is hard to say how the object first seen by the Magi in their eastern skies was the same object they saw perhaps weeks or months later when traveling to Bethlehem. Moreover, because meteors are so short-lived, it is hard to talk about how it could have risen in the east at all. Considering how many read the phrase about the Star rising as being connected to sunrise, a meteor then in the east would have been invisible, drowned out by the light of dawn. Also because of their fleeting perceptibility, it is hard to see how a meteor can last long enough to "go before" the Magi during the full trip to Bethlehem from Jerusalem, which on foot would probably take on the order of two or three hours. While a series of fortuitous meteors all pointing towards the south could do this, that scenario would not fit with the singular Star mentioned by Matthew. The astronomical difficulties do not end there. We cannot see how a meteor can be said to arrive and stand over a particular location, a point that Moore fails to overcome. Such a moving flash of light could not fit this bill.

The last detail to point out is that there is simply no reason to think that a meteor would have been particularly auspicious and pointed to the birth of anyone. Meteors, as noted above, are extremely common. They are seen throughout the sky and can be seen on most any given night so long as there is good visibility. In a shower, meteors can burn up in the sky at a rate of more than one per minute. If meteors were so auspicious, one would think that magi would be going all over the world looking for kings nearly constantly. However, we have no record to support this absurd situation even partially. The only potential solution is if there were other things in the sky that had the needed astrological import. That discussion will come later in Chapter 5. In the meantime, it is safe to leave aside this proposal as there are too many astronomical and symbolic problems.

However, there is a more prominent transient object that we know gets the attention of people: the comet. These dirty snowballs produce a tail of gas and dust when they approach the Sun; the heat causes the comet to evaporate at the surface, and the tail of the comet can be millions of kilometers long, much farther than the distance from the Earth to the Moon. A comet, such as the famous one named after Edmund Halley, will usually be visible to the naked eye for days or weeks, perhaps even months. Because of their prominence, we are fortunate that many ancient cultures, from China to Mesoamerica, record their entry into the sky and what they meant to them. This is already significantly better than the meteor hypothesis because we can know if there was such an object in the sky, where it was, and we can investigate what they were thought to mean since we have these ancient eye-witness records of comets providing the time they were seen, where in the sky they were seen, and often providing descriptions of the comet itself (and interpretations of comets are numerous).

As for the astronomical circumstances that could be consistent with Matthew's account, it is relatively easy for a comet to be said to rise in the east. Depending on its ephemeris (the path of the object in the sky), a comet could appear on the horizon at a time that could have greater astrological significance. Considering that many interpret the phrase "at the rising/in the east" (*en te anatole*, ἐν τῇ ἀνατολῇ) from Matt 2:2 to mean the Star rose at sunrise, a comet can also fit this bill. Records of comets from the past often do not have enough details to make such an ephemeris impossible, so a comet can be consistent with most any reading of Matt 2:2.

Moving to verse 9, a comet can be seen towards the south, so it could be in the direction of Bethlehem from Jerusalem when the Magi headed out. As for the Star standing over where Jesus was, the two references which have comets standing over places that proponents point to are from the Roman historian Cassius Dio[33] and the Jewish historian Flavius Josephus.[34] So far, it seems we can have a match between the astronomical aspects of a comet and Matthew's account.

Lastly, we need to consider if a comet was seen in the timeframe of Jesus's birth according to Matthew's chronology. At this point we are fortunate because of the excellent records of comets from China. These Far Eastern reports not only tell us when and where a comet was seen, but it even gives details of how the comet looked. As we have it, in 5 BCE there was a comet with a notable tail which the Chinese called a *hui* comet or 'broom star'; another tailless comet was seen in 4 BCE but after Passover and the death of King Herod. Before the comet of 5 BCE is Halley's Comet in 12 BCE, but this

---

[33] Cassius Dio, *Roman History* 54.29.8.
[34] Josephus, *Wars of the Jews* 6.289f.

is likely to be too early to fit.[35] Nonetheless, we have a reasonably good candidate for a comet that fits the timeline.

While this is also superior to the meteor hypothesis, the comet still has significant problems fitting the description of Matthew's account. For example, while a comet would potentially be seen in the direction of Bethlehem from Jerusalem as argued by the material scientist Colin Humphreys,[36] the text says that the Star moved in that direction—the Star "went before" the Magi. The movement of the comet would only be able to perceptively travel from east to west, not from north to south. Moreover, a comet cannot arrive and stop as the Star is said to have done, let alone hang over a particular hamlet. The examples mentioned above of comets standing over places do not actually fit when one looks at the original text rather than an English translation. In the original Greek they use significantly different words than does Matthew and none have a comet initially moving and then stopping. While the Evangelist uses the term *histemi* (ἵστημι) to mean stand, *erchomai* (ἔρχομαι) for arriving, and *epano* (ἐπάνω) to mean over, Dio uses *aioretheis* (αιωρηθεὶς) that means 'to hang' rather than stand and *huper* (ὑπέρ) to mean over. Similarly, Josephus uses the preposition *huper* instead of *epano*. The difference between *huper* and *epano* will be discussed in Chapter 7, but in the meantime it can be said that these historians mean something significantly different than what Matthew describes. But even

---

[35] Ho Peng Yoke. "Ancient and Medieval Observations of Comets and Novae in Chinese Sources", *Vistas in Astronomy*, vol. 5 (1962): 127-255, esp. p. 137.
[36] Colin Humphreys, "The Star of Bethlehem—a Comet in 5 BC—and the Date of the Birth of the Christ", *Quarterly Journal of the Royal Astronomical Society*, 32 (1991): 389-407.

excusing the vocabulary differences, neither of these authors has a comet standing over a particular house which Matthew implies, which again will be specified in Chapter 7. This means there is no precedent in the literature to say a comet can stand over a location, let alone travel south and stop—another claim to which no parallel is provided.

But perhaps the most obvious problem with the use of a comet is that it is one of the few astrological signs that are almost universally agreed upon: they are a terrible sign! The late astronomer Carl Sagan and his wife and TV producer Ann Druyan in their book *Comet* describe how from China to the Mediterranean to Mexico comets were so often seen as signs of evil. Dozens of examples can be given, but let us see a few in the regions that are of interest to this study. The writer Diodorus Siculus described a comet which appeared to have three tails that was seen as a sign of the end of the powers of the peoples of the southern peninsula of Greece, the Lacedaemonians, better known as the Spartans.[37] A nephew of the Roman philosopher Seneca the Younger wrote the following: "The heavens appeared on fire, flaming torches traversed in all directions the depths of space; a comet, that fearful star with overthrows the power on the Earth, showed its horrid hair."[38] Seneca would seem to agree with his nephew in his own writings about comets.[39] The astronomer and astrologer Ptolemy states that comets are a sign of wars, hot weather, and disturbed conditions,[40] all of which is consistent with the

---

[37] Diodorus Siculus, *Biblotheca Historica* 15.50.
[38] Carl Sagan and Ann Druyan, *Comet*, p. 25.
[39] Seneca, *Naturales Quaestiones* 7, especially sections 1.5; 17.3; 28:2-3.
[40] Ptolemy, *Tetrabiblos* 2.9 (90f).

astrologer Manilius's opinion of the matter.[41] Even Humphrey's own examples of comets that stood over things in the historical record were seen as a both a sign of doom for the Jews during the revolt[42] and a portent at the death of Augustus's friend Agrippa.[43] On top of these, ancient Western sources could add dozens of cases with the same explications.

Now, all of these are examples among Greeks and Romans. But the situation is the same in the Middle East. Cuneiform texts concerned with celestial signs, including letters by astrologers, give connections between a comet and the death of a king.[44] The death of a king is hardly the birth of one, so there is no hope of twisting this omen into a natal premonition, let alone a favorable one. In Zoroastrian literature, there is an object known as Mush Parik, and this was supposed to be a comet occasionally released from the Sun. It was a terrible sign and foretold of abysmal events.[45] Later, after the Arab conquests, the records of the Persians and Greeks would be read and pondered by Muslims and other Arab-speaking scholars. Their records indicate that comets were also ominous and ruined a horoscope.[46] From the whole of antiquity and into the medieval period, I have found in Middle Eastern records no

---

[41] Manilius, *Astronomica* 1.874-926.
[42] Josephus, *Wars of the Jews* 6.314-315.
[43] Dio, *Roman History* 54.29.7-30.1.
[44] Hermann Hunger, F. Richard Stephenson, C. B. F. Walker, et al., *Halley's Comet in History*, p. 18; Gary W. Kronk, *Cometography: A Catalog of Comets*, p. 1; Hermann Hunger, *Astrological Reports* §§ 339, 456.
[45] *Greater Bundahishn* 5.4-5; 5 A.5-7; *Shkand gumanig wizar* 4.46. See *Encyclopedia Iranica*, vol. 2, p. 867.
[46] Edward S. Kennedy, "Comets in Islamic Astronomy and Astrology", *Journal of Near Eastern Studies* 16, 1 (1957): 44-51.

exception to the rule that comets were signs of evil to come.

This problem is well-known to everyone that argues in favor of a comet as the Star of Bethlehem, so there is a heavy reliance on the very few exceptions in the literature from antiquity. The most prominent counter concerns the comets seen at the birth and ascension of King Mithridates of Pontus, a ruler in modern-day Turkey who had been a significant opponent to the Romans in Asia Minor in the first/second century BCE. The timing of these comets was at one time considered a fiction, but they seem to be confirmed in Chinese annals and coins. Unlike other despots that would avoid the use of such an ominous symbol, Mithridates had the comet on his coinage (a trait he would share with Vlad the Impaler), though only on the smallest of coin denominations.[47] However, it seems that the way the symbol was used was done *post factum*, that is, after Mithridates seized power, and the comet required reinterpretation. The coins do not simply show the comet but couch it in other astrological symbols, such as the constellation Pegasus, so it seems that the ill signs were being mitigated, as suggested by classicist John Ramsey. The comet is interpreted as a great flame rather than a comet straight-out and thus conforming to Persian prophecies.[48]

The same situation happened with the Octavius's (later known as Caesar Augustus) comet; it was transformed into a symbol of power by mythologizing it into Julius Caesar's soul rising to heaven. This was also

---

[47] John Ramsey, "Mithridates, the Banner of Ch'ih-yu, and the Comet Coin", *Harvard Studies in Classical Philology* 99 (1999): 197-253.
[48] Ramsey, *op. cit.*, pp. 228-229.

possible because when the comet was first seen the tail was not noticeable, thus appearing as a star rather than a comet. It would be Augustus's enemies that would call the new object a comet and a sign of the coming civil war, while his allies more often referred to it as a star (*sidus*).[49] Thus, the identification and interpretation was politicized, and when viewed as a comet by Octavian's detractors the comet was given its evil classification, thus adding another example of negative interpretations of comets.

In these cases, we see that the comets were not viewed as positive omens in their own right but were construed by those in a position of power to influence how these objects would be seen after the fact as propaganda. That is completely different than what is needed in the case of the Star of Bethlehem: the Magi see the comet and know it is an auspicious sign for another nation's king. They do not know before seeing a comet that some newborn is their political master, as was the case of Mithridates, so they had no reason to try and reimagine how the comet was in fact an excellent sign and not really a nefarious comet. The interpretation by the Magi is not *post factum* in favor of a major political figure, so the examples of Mithridates and Augustus do not fit what is needed to conform to Matthew's story. But even if we ignore how the symbol was being drained of its harsh portents after the fact by a particular political figure for his own benefit, this is still an exception (two at best) in a list of many dozens that are distinctly negative. We still have to say that comets were seen as ominous signs in well over 95% of the cases. Without showing how the comet of 5 BCE could be made out to

---

[49] John T. Ramsey and A. Lewis Licht, *The Comet of 44 B.C. and Caesar's Funeral Games*, pp. 135-153.

be a great omen rather than an ill one via some other astrological circumstance, then we have every reason to discount the comet hypothesis based on its symbolism alone. Combined with its failure to fit the description of Matthew's Star, the comet hypothesis is a failed one.

However, there exists one potential trick up the sleeve of the adherent to the comet hypothesis: the testimony of Origen, perhaps the most scholastic Christian in the first few centuries of the Common Era. In his point-for-point response to the attack on his faith by the pagan philosopher Celsus, Origen talked about the Star of Bethlehem and comets.[50] Origen says how there are plenty of examples where a comet was seen and it wasn't that bad, so having the Star of Bethlehem being like a comet was not a poor heavenly sign. Many take this to mean that Origen thought the Star was a comet.

However, Origen only says the Star is like a comet in how it has significant meaning and how it is newly seen in the sky, unlike planets or other stars. Moreover, Origen does much to contradict the view that it was a comet that made people think a king was born. He said that he knew of no prophecy about a comet being seen at a certain time and heralding the birth of a king of any region. Considering how bookish he was, that suggests that there was no such connection in the ancient world— no one was expecting a comet to prophesy a king. Furthermore, Origen in other works shows the Star was a miraculous object that came down to Jesus just like the dove that perched on him at his baptism.[51] Origen did not think the Star of Bethlehem was a comet, but instead he brought up the tailed star as a point about

---

[50] Origen, *Contra Celsum* 1.58-59.
[51] Origen, *Homilia in Numeros* 18.4.

momentous astral symbols at the start of major events. Origen leaves the comet hypothesis for the Star with nothing as he sides with a supernatural understanding of the Star, and the significant problems with the hypothesis, mentioned here, make it untenable.

Chapter 4: Novae and Supernova

This chapter focuses on what is one of the newer yet popular hypotheses concerning the Star of Bethlehem, and this newness is because for so long the phenomena concerned were not well-understood. Here we want to look at novae and supernovae. These are different objects, but they are effectively the same for naked-eye observers: a bright light which was called a new star due to its sudden appearance, hence the name of 'nova'. The mechanisms for novae and supernovae are different, and there are different versions of each type, but they are both basically a stellar explosion. In the case of a nova, an object known as a white dwarf accumulates hydrogen on its surface from a near-by star; when there is enough gas there, nuclear fusion can take place and make a burst of light. For supernovae, there are two basic types, ingeniously named Type Ia and Type II. Type Ia are similar to novae in that they involve gas being accumulated by white dwarves, but when the mass of the white dwarf is large enough it collapse and releases a huge amount of energy. In the case of Type II, a giant star, far larger than the Sun, has exhausted its nuclear fuel, collapses due to gravity, and then the in-falling gas slams into the core of the star, bouncing and producing some of the largest explosions in the universe.

If bright and close enough, novae and supernovae will appear in the sky to the naked-eye observer as a bright point of light, not unlike a normal star. What will give them away is their recent appearance and how they slowly dim away. A nova may lose its brilliance in a matter of days, and a supernova may still be visible for months; a year after seeing a supernova, there is unlikely anything to be seen without a telescope. Novae can be recurrent should they still have a continuous source of hydrogen gas, but supernovae are one-time events of brilliant power. A supernova can have the luminosity of an entire galaxy of stars. Unfortunately for us in the 21st century, there has not been a visible supernova in our galaxy in a long time, the last one being witnessed by Johannes Kepler in 1604. He was also fortunate to see one in 1572, while we have had a dry spell of such explosions for centuries. The supernova in 1987 from the Large Magellanic Cloud was for a time visible, but it was outside our galaxy and not nearly as brilliant as Kepler's star. Many think one is long overdue, and a red giant star like Betelgeuse in the constellation Orion could go at any time.

The question then is if such an object could be what was seen by the Magi 2000 years ago. A brilliant light in the sky captures the imagination, not to mention looks well on a Christmas card, so perhaps it makes sense that this is one of the rather popular conceptions of what the Star could have been. The most admired incarnation may be from the award-winning short story by Arthur C. Clarke,[52] one of the top science fiction authors of the 20th century, especially known for writing *2001: A Space Odyssey*. In his story, a Jesuit astrophysicist comes to a distant planet to discover the remains of a lost and prosperous civilization that was destroyed when the sun of that world had gone nova. After some calculations, he determines that this star that exploded would have its extreme luminescence reach Earth at the time of Jesus's birth. This drove the Jesuit astronomer into a crisis of

---

[52] Arthur C. Clarke, "The Star", *Infinity Science Fiction* 1, 1 (1955): 120–127.

faith, asking how his good God could destroy a civilization for the sake of making the Star of Bethlehem. The story won the Hugo award and would later be adapted into an episode of the 1980s reboot of *The Twilight Zone*.

Besides fictional and popular conceptions, the idea of a supernova as the Star was put forward by professional astronomers in the peer-review literature in 1977[53] and was also supported recently by the astronomer Mark Kidger.[54] The theoretical physicist Frank Tipler also has his own version of the supernova hypothesis.[55] What features do they see that can explain the Christmas Star? Firstly, the newness of the nova would be spectacular to those familiar with the sky; a new object could well fit what Matthew meant by "his Star" to distinguish it from all the other stars and planets in the sky. Like a star, it can rise in the east, and again it can fit the meaning of "at the rising" from Matt 2:2 just as well as a comet. Also like a comet, the nova could be in the direction of Bethlehem from Jerusalem, which is then seen to be before the Magi. On the other hand, Tipler believes the supernova was at the zenith so it could stand over Bethlehem, fulfilling a different aspect of the description of the Star. In fact, Tipler never even attempts to explain the movement of the Star, leaving his hypothesis weakened even before critical analysis.

As for what the nova would have meant to the magi, unlike the case for comets, we have effectively nothing to work with in knowing how novae were interpreted

---

[53] David H. Clark, John H. Parkinson, and F. Richard Stephenson, "An Astronomical Re-Appraisal of the Star of Bethlehem—A Nova in 5 BC", *Quarterly Journal of the Royal Astronomical Society* 18 (1977): 443-449.
[54] Kidger, *The Star of Bethlehem: An Astronomer's View*
[55] Frank Tipler, "The Star of Bethlehem: a Type Ia/Ic Supernova in the Andromeda Galaxy?" *The Observatory* 125 (2005): 168-174.

astrologically. The divinatory treatises of Ptolemy and others say nothing about them, and records of novae in ancient Greece, Rome, and the Middle East are very sparse. One of the few records of a nova by someone from this time and region comes from Hipparchus of Rhodes in the second century BCE,[56] but we have no interpretation of the object besides how it inspired Hipparchus to make a careful map of all the stars to see how much the sky changed over time, countering Aristotle's notion that the heavens were unchanging.

However, if Chinese records are any indication, a nova may have been indistinguishable from a comet. In the catalogue of Chinese comet reports, three categories are noted: *hui* or "broom star" comets, *po* or "shining" comets, and 'guests stars'. The latter are considered novae, while the first in the list are certainly comets as they have tails. The *po* comet, on the other hand, is ambiguous; it could be a comet that has its tail pointing directly towards or away from the Earth so it looks like a singular object like a nova. Some *po* comets today are thought to be novae. So it may be the case that novae would have been thought to be a comet in the West; even if distinguished by European or Babylonian astronomers from comets, it may have been seen as similar in meaning to a comet because of their sudden appearance. At best we cannot know what the ancient magi would have thought of a nova, and at worst it would have been seen in the same way as a comet, hence a negative rather than positive omen (see Chapter 3).

Before considering how well novae actually do fit the details of Matthew's account, we need to see if any such object was observed in the timeframe of when Jesus may

---

[56] Pliny, *Natural History* 2.96.

have been born. The Chinese records as put together by Ho Peng Yoke show no 'guest stars' anywhere near the timeframe of 10-4 BCE, but there is a *po* comet recorded in 4 BCE. This *po* comet comes too late in the year to have happened before Herod's death (no later than early 4 BCE), so it won't fit the timeframe. The comet in 5 BCE is said to be a *hui* comet and with a tail, so it should not even be a potential nova/supernova candidate. That is, unless one believes Kidger and the researchers he follows. He believes that the object in 5 BCE and the other in 4 BCE are the same object; there was a mistake in the date of the record and in fact it was a nova in 5 BCE.

This should be automatically a suspect argument for several reasons. The first point to consider is the distance the objects are from each other in the sky. Treating the stars as lying on a sphere, astronomers talk about the angular distance two objects are from each other; like lines of longitude, there are 360 degrees to go around in a circle, and difference in longitude indicate angular distance. A good rule of thumb is that the Moon is about half of a degree across as seen from the surface of the Earth; your thumb when held out at arm's length will cover the moon and is about 2 degrees wide, making the phrase 'rule of thumb' very apt in this circumstance. The 4 BCE object was 20 degrees (40 moon diameters) away from the location of the 5 BCE object, which is simply too amazingly large of a mistake in placement. The 5 BCE object was away from the plane of the galaxy (that is, what is commonly seen in the sky and called the Milky Way) in which the vast majority of the stars are located, making it less likely a nova; with far fewer stars, especially massive ones, there are fewer opportunities to have an exploding star. Lastly, the 5 BCE object is said to have had a tail, so it cannot be a nova which do not have tails. If the astronomers making this argument were correct, it means that the ancient eastern astronomers got the time, location, and identification of the object wrong. If so, then why trust them for any of this data? But this is only worth considering if the astronomers actually had a good reason to think these things.

However, the only premise that this strange line of inquiry stood upon was that the Chinese did not record the 4 BCE object, while the Koreans did, though with a nonsensical date (a sort of February 31st); since the Chinese comet records are far superior in quantity and quality, it seems likely that the Korean record was a mistaken duplicate of the 5 BCE comet/nova. All of this was soundly refuted back in 1979 by Chinese historian Christopher Cullen who showed that the reasons given were completely wrong; the error was only in a Korean copy of the original *4 BCE* Chinese record which we also have but without the date slip-up. The Chinese did record the 4 BCE comet (which medieval Korean scholars miscopied), and it was mentioned on the very page of the list from Ho Peng Yoke's compilation of Chinese comet records Kidger and others used to make their argument (the same paragraph, no less).[57] Kidger and those he follows have produced nothing to contradict Cullen, and they cannot because their argument is based on the premise that there is no Chinese record of the comet in 4 BCE, which is false. What this eventually means is that there was no observed nova for the time of Jesus's birth.

The hypothesis of Tipler tallies even less with the historical record. His supernova in the Andromeda

---

[57] Christopher Cullen, "Can We Find the Star of Bethlehem in Far Eastern Records?" *Quarterly Journal of the Royal Astronomical Society* 20 (1979): 153-159.

galaxy has not even the 5 BCE comet to connect it, and it is unlikely it was even visible. Based on current physics models of supernovae, Tipler calculated that the maximum theoretical brightness of such an explosion could be observable at the naked-eye level (though he underestimated the distance of Andromeda to us), but the brightness is even less than the apparent brightness of the galaxy itself, and that assumes no dust or gas is blocking the light of the object (a very unsafe assumption); it would not register for someone without a telescope. We should also realize that the Andromeda galaxy somehow went unobserved until it was sighted in the Arab-speaking world in the 10th century,[58] hardly then a place anyone was looking or finding any supernovae. Considering no supernova has ever been seen in Andromeda without the telescope, and only one has been witnessed with such a tool (in 1885), and while such explosions are supposed to happen on the order of once per century, then the complete lack of records of sighting of the galaxy and any explosions in it proves no one was noticing or could notice anything happening in that spot in the sky. We can be reasonably certain no supernova could have been nor was seen between 10 and 4 BCE.

Ignoring the problems of interpretation, and if such an object was actually seen at the time of Jesus, could the other criteria of a nova or supernova really fit what Matthew says? Consider again the point of the Star "going before" the Magi. Unlike a comet, a nova cannot even move relative to the stars in the sky, and all stars move from east to west, not north to south as Matthew implies when he says the Star "went before them", and the Magi are traveling south to Bethlehem from

---

[58] George Robert Kepple and Glen W. Sanner, *The Night Sky Observer's Guide,* Vol. 1, p. 18.

Jerusalem. Again, the Star 'goes', not 'is' before the Magi, so motion is implicit and must travel in the same direction as the path of travel to Bethlehem. The notion of the Star standing over where baby Jesus was is also not easily explained. On Tipler's hypothesis of the nova at the zenith, it would no more be over Bethlehem than over Jerusalem without extremely accurate measuring equipment unheard of in the first century. Kidger, on the other hand, just says the Star was high in the sky towards the south.[59] But this fails to explain the implicit motion of the Star, the fact that Matthew says it stops, and how it can stand over anything if, on Kidger's reading, it's not over the town at all, let alone a particular house, when the Magi are there and the Star is still south of them.

While a supernova at the birth of Jesus would have been spectacular, it simply doesn't fit what Matthew describes. What is equally important is that no such event can be shown to have happened that fits the timeline, and there is no evidence to show it would be associated with the birth of a king, Jewish or otherwise.

---

[59] Kidger, *The Star of Bethlehem: An Astronomer's View*, p. 35.

Chapter 5: The Planets and their Positions

Perhaps the most popular of the hypotheses for the Star of Bethlehem have dealt with the planets, their motions against the background stars, and their meaning to the astrologically minded. In the literature there are copious musings about the planets Jupiter and Saturn, the constellations Pisces and Aries, and individual stars such as Regulus in Leo. Many attempted interpretations of these objects have created a mass of confused and conflicting thoughts on the subject, but a few common and basic ideas will come to the surface in this analysis. The motions of the planets have also persistently been correlated with the phrases used by the author Matthew, especially how the Star "went before" and "stood over".

There are numerous versions of this attempt to explain the Star that I will split into two larger groups: a conjunction of two planets (i.e. Jupiter-Saturn, Jupiter-Venus), and the horoscope in its entirety (that is, considering all the planets and the location of the cardinal points). There are fewer scholars in the last camp, but it also includes one of the most extensive attempts to connect the Star, Matthew's account, and history. There is significant overlap between the hypotheses as well, so it is worth reading both sections to understand the advantages and disadvantages of a general hypothesis that uses the planets to elucidate the Star of Bethlehem.

Part A: Conjunctions

First, a point of definition is needed since this is not usually understood correctly. In the system of stellar coordinates, there are the equivalents of latitude and longitude called declination and right ascension. Declination, like latitude, measures from what is called the celestial equator, which is simply the Earth's equator projected out into space. Right ascension measures along the celestial equator similar to longitude. Using this system, a conjunction means that two planets orbiting the Sun (or a star and a planet) seen from Earth have the same right ascension in the sky; basically, they are on the same line of longitude. It does not mean that the two objects are necessarily close to each other, but when it comes to the planets this is more often the case. Because all the planets are about in the same plane as the Earth about the Sun, they will have minimal latitude differences and so at a conjunction they can be close. Sometimes planets can be so close that one can cover the other, though this is very rare; more common is an object such as the Moon blocking a star or planet which is called an occultation. These points are highlighted here so that way a thesis about the Star is not dismissed simply because the planets are not close enough; closeness is not a criterion of conjunctions, though it may be for some arguments about the Star. The closeness of the conjunction should only matter if a proposed hypothesis must have the objects so close that they are indistinguishable and appear as a single object (as some have argued).

When it comes to the conjunction hypothesis, there are two major contenders in the literature and in planetarium shows. The first was the one connected to Johannes Kepler, that of the conjunction of Jupiter and Saturn in 7 BCE (actually, there were three in that year), and it has the longest and most interesting history. The second hypothesis concerns Jupiter and Venus, along with the star Regulus; this idea came into notice first in the 1960s and then greater mainstream prominence in the 1980s. The latter hypothesis will be considered first, and more attention will be given to the Jupiter-Saturn conjunction(s) since it has had the greater consideration (including a partial endorsement by the Benedict XVI, Pope emeritus)[60] and diversity of thought.

The conjunction of Venus and Jupiter was first brought to light by Robert Sinnott,[61] a writer for the popular astronomy magazine *Sky & Telescope*, but the first person to consider it in depth was a meteorologist and follower of the evangelical Hebert Armstrong, Ernest Martin.[62] The events in question take place between 3 and 2 BCE, which is outside the allowed time period for Jesus's birth; because of this, Martin's writing on the subject is quantitatively more on arguing for a different chronology in which King Herod dies (not before the Passover in 4 BCE but in 1 BCE). So Matthew's story is historical, so long as you change history! The arguments of Martin and others will not be considered here as they have not been convincing to historians.

Nonetheless, let us suppose the dating of Herod's death is not well-known. Ignoring the chronological issue, what was it that was so important about this

---

[60] Joseph Ratzinger, *Jesus of Nazareth: The Infancy Narratives*, pp. 97-101.
[61] Robert Sinnott, "Thoughts on the Star of Bethlehem", *Sky & Telescope* 36 (1968): 384–386.
[62] Ernest Martin, *The Star that Astonished the World*.

conjunction? What is interesting is that it combined two symbols in the sky with kingship: Jupiter, the king of the gods, and the star Regulus, a name derivative from the Latin for king (*rex, regis*) found in the constellation Leo, which Martin argues was associated with the Jews. After a conjunction with Venus, Jupiter hangs about Regulus for several months, appearing to go backward and forward around the star, perhaps highlighting what is to come. Then, in 2 BCE, Venus and Jupiter had such a close conjunction that they could have appeared to be a single object. With some effort, evidence from the Bible is used to show that Leo is the constellation associated with the Jews, making conjunctions there important for that people or their religion. The first conjunction is also visible on the eastern horizon before sunrise, thus in line with the Star seen "at rising". As for the business of the Star "going before" the Magi, Martin either says it was symbolic or that Jupiter alone was seen southward, similar to what has been said of comets and novae in the previous chapters. As for the Star standing, this is in reference to Jupiter reaching the apparent stationary point relative to the other background stars in the sky exactly on December 25, 2 BCE. Merry Christmas!

Because the point of a planet being referred to as stationary or in retrograde will come up several times more, let me explain what is meant here. To observers on Earth, the planets of the solar system are star-like objects that appear to move against the background of the other stars; the name 'planet' comes from the Greek for wanderer as they appeared to not just move but to move in strange ways. Usually the planets will travel westward relative to the background constellations, but occasionally they slow down, stop, travel backward to the east (what is called retrograde motion) for a bit and make a loop, stop again, and finally travel westward again. This odd motion we understand as due to an optical illusion because the planets go about the Sun at different rates and the Earth passes them from time to time. It is a similar observation when you overtake another car; at first the slower car is ahead and appears to move forward, but then as you begin to pass it seems to go backward for a bit. In ancient times, this was a significant mystery and one of the popular theories of the time was used by Ptolemy; this was an earth-centered (geocentric) system (actually, Ptolemy set the Earth a little bit off from the center of motion of the planets), and the planets went about on circles, but not so straightforwardly. In fact, the planets stayed on rotating circles whose centers went about in a circle, creating what is called an epicycle—a cycle on top of another cycle. As such, sometimes the planet's overall motion was backward compared to the motion of the larger cycle.

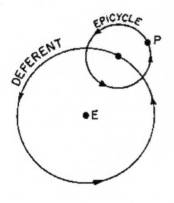

The basic geocentric model of Ptolemy. The Earth (E) is at the center of the deferent circle, while the Planet (P) rotates about on the epicycle. The center of the epicycle travels along the deferent circle. The combined motions of the planet on the epicycle and the movement of the epicycle give the apparent backward motion of the planet in the sky to earth-based observers every once in a while. This model of retrograde motion, with other complications such as the equant, lasted for centuries, though modified by Arab astronomers and finally abandoned during the Scientific Revolution.

This sort of system was apparently used in the ancient computer known as the Antikythera mechanism which used the eponymous epicyclic gearing to model planetary motion. While strange-sounding to moderns, the model was very successful in its day in predicting the future location of the planets, and even the solar system of Copernicus with the Sun in the center (heliocentric) was not superior.

This whole business is important to various Star researchers because they will talk about the retrograde motion of the planets as related to the "going before" of the Star, and the stationary point as related to it standing still. While Martin in his Jupiter-Venus hypothesis does not talk about retrograde motion in this way, others do, and more discussion will concern these.

With this preliminary background out of the way, let us consider how well the thesis actually fits the description of Matthew's Star. On the point of the Star "rising", there is nothing apparently at fault with the hypothesis. However, the text specifies a single Star, not two as Martin has it with the close conjunction of Jupiter with Venus. Perhaps it was Jupiter that was the important star, as others theorize, so let us accept this and simply move on to more important aspects. When it comes to the motion of the Star, Martin's explanation fails in the same way as it does for novae and comets. Jupiter does not travel north to south, nor does any other planet; moreover, in Martin's hypothesis, Jupiter is at a stationary point and not moving at all with respect to the stars. As for the condition of stopping, a stationary point is not something that can be seen happening in an instant, but instead a planet will appear to be stopped for days. Matthew talks about the Star going before and then stopping, a sequence of events taking place in the couple hour walk to the small town where Jesus was; this specifies that the stopping happened in the short time it takes to travel from Jerusalem to Bethlehem, probably even less. Moreover, the text of Matthew uses a verb form (discussed in Chapter 7) that specifies an instantaneous stopping action, not one that could be spread out for days.

We also have no good explanation as to how the Star is "over where the young child was" given this hypothesis. In Martin's reading, Jupiter happened to be over Bethlehem as seen from Jerusalem just before dawn, and that, argued Martin, is what Matthew meant. However, the text says that the Star came to stop over Bethlehem *after* it "went before" the Magi. For Martin, the events are happening at the same time—going before and standing over. These difficulties will be further explored more fully in Chapter 7, but the key word that upsets Martin's hypothesis is 'until': the Star went before the Magi "until" it stood. Even without considerations of the Greek, we can see in English that the events of the Star moving and stopping are not simultaneous, and the stopping action took place after the movement of the Star.

But what about the possible symbolic importance of the events in the sky? Unfortunately, Martin gives no ancient evidence that a Jupiter-Venus conjunction would specify the birth of anyone; instead, cuneiform literature says that such a conjunction was a sign of war and hostilities toward the king, not the birth of one.[63] Nor is there evidence that Jupiter retrograding about Regulus had any positive meaning for eastern astrologers. Martin only has speculations. Had he actually looked into

---

[63] Hermann Hunger, *Astrological Reports* §§ 212, 448.

eastern records, he would have found that this very retrograde motion about Regulus by Jupiter was a sign of someone overthrowing the king, a terrible omen, rather than the birth of one.[64] If anything, the Magi should have been worrying about a civil war in their own land, not a baby in Judea.

As for the constellation Leo being the sign of the Jews, Martin looks to two verses from the Bible, Genesis 49:9 ("You are a lion's cub, O Judah; you return from the prey, my son. Like a lion he crouches and lies down, like a lioness—who dares to rouse him?"), and Revelation 5:5 ("Then one of the elders said to me, 'Do not weep! See, the Lion of the tribe of Judah, the Root of David, has triumphed. He is able to open the scroll and its seven seals.'"). Unfortunately, this has three obvious problems. One is that neither of these verses is about the stars or constellations, so the connection between Judea being like a lion (and what nation *doesn't* want to have such a symbol for their own?) is not the same as Leo representing the nation. The second is that there is no reason why eastern Magi would be using the Bible for their astrological geography. Instead of going to an actual magian text, selective exegesis of the Bible is used to fill in the gap, but that is pointless unless it can be shown the Magi would have been doing the same thing. No evidence exists, leaving a connection between Leo and the Jews by the Persians as speculation, not fact. A third point worth mentioning is that we could have done this same exegesis for most any constellation we wanted to. Revelation 5:6 says Jesus is the Lamb, so perhaps the constellation of choice should be Aries the Ram? To see how suspect Martin's exegesis is, one can turn to the

---

[64] Hunger, *Astrological Reports* § 279.

Talmud for a debate about astrology and Judaism. According to Rabbi Johanan, there is no *mazal* (astrological fortune, meaning, in the context, no planet or constellation) for Israel, but only for other nations that do believe in such things.[65] One cannot say that some particular star sign represents the Jews or the Jewish religion when we have such testimony denying any such connections. When European rabbis in the Middle Ages took up astrology, they also chose different constellations than Leo, such as Aquarius.[66] This will be discussed more below. As such, even if the Bible were used by the Magi to determine a constellation representative of Israel, there is nothing that shows Leo was the preferred interpretation.

Martin's entire case is constructed on a series of baseless speculations, many of which can be shown to be false. It doesn't help that the events are also outside of the plausible timeframe for Jesus's birth and do not fit Matthew's description of events. The problems with Martin's description have not even been fully explored, so how far off he is from actually explaining the Star will be shown even more starkly in Chapter 7.

Assuming the Jupiter-Venus conjunction does not fit the bill for astrological auspiciousness, it is now time to consider the longest-running conjunction connection to Jesus. While many associate it with Kepler, the idea actually goes back centuries before his time and to a different land. The thesis actually rests on the idea that the periodic conjunctions of Jupiter and Saturn, about 20 years apart, have an important effect on world affairs. Those effects are strongest when the conjunctions are

---

[65] Babylonian Talmud *Shabbat* 156a.
[66] Bernard R. Goldstein and David E. Pingree, *Levi ben Gerson's Prognostication for the Conjucntion of 1345*, pp. 35, 37.

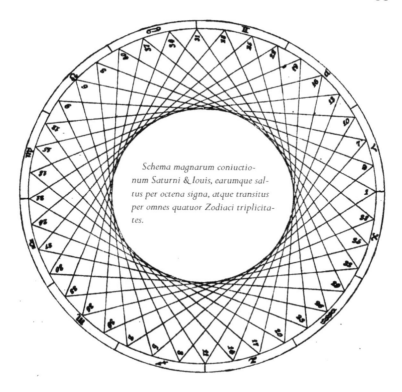

A zodiac map of the successive conjunctions of Jupiter and Saturn, starting at the first degree of Aries. Each conjunction is numbered up to 40, after which the conjunctions repeat. This image is reproduced from Johannes Kepler's Mysterium Cosmographicum (1596), and it was this astrological diagram that led to Kepler's first (though erroneous) epiphany into the structure of the solar system. The text in English reads "A schema of the great conjunctions of Saturn and Jupiter, and their leaps through eight signs, and crossing through all four quartiles of the Zodiac."

first seen in a new set of three constellations called a trine. Each trine is associated with one of the ancient four elements of earth, wind, fire, and water (and unfortunately not with heart), and the Jupiter-Saturn conjunctions tend to bounce around in a single trine for decades. When a conjunction happens for the first time in the Fiery Trine (Aries, Leo, and Sagittarius) this is the most powerful, and this happens only about every 960 years (or 800 when considering the effects of precession). Because such a cycle is nearly a thousand years, this system of astrology (the great conjunctions) became particularly interesting to Zoroastrians, the religion of the magi, who had an affinity for millennialism.

Does this mean then we are in luck? Can a great conjunction actually be said to have interested the Magi? Unfortunately, there are things we know about this hypothesis, and part of that is when this practice began. While it did originate in Persian territories, it did not happen until the Sassanid period in Persia which began in the third century CE. Then, kings such as Shapur I would send out scholars to gather up the scientific knowledge of the Greeks, Romans, and Indians; afterward, they used their astronomical and astrological works to create the great conjunction system.[67] At best, it comes several centuries too late. We also don't see this system applied to individuals until the rise of Islam.

The first time the great conjunction was connected to Jesus appears to be with the Jewish Persian astrologer Masha'allah in the 8th century, whom we know about concerning this subject from the 9th century Christian

---

[67] David Pingree, "Astronomy and Astrology in India and Iran", *Isis* 54, 2 (1963): 229-246; "Historical Horoscopes", *Journal of the American Oriental Society* 82, 4 (Oct-Dec 1962): 487-502.

astrologer Ibn Hibinta.[68] However, when Masha'allah did use this method, his timeframe was far different than what any modern researcher believes fits the lifetime of Jesus, and this is because he followed the method correctly and later scholars researching Jesus and his birth do not. As mentioned above, the most important conjunctions of Jupiter and Saturn take place in the Fiery Trine for the first time, and this happened in Leo in 26 BCE. According to our source, Masha'allah said Jesus was born in 13 BCE, though he also gives the more traditional date of 2 BCE.[69] Later on, a Muslim astrologer famous in medieval Europe, Abu Ma'shar, would fit a great conjunction closer to the time of Jesus's birth, the now-famous one in 7 BCE. This idea would later travel to Europe, and it would be the ultimate source for Kepler's speculations on the subject.[70]

So while the idea of a great conjunction has considerable antiquity in its association with Jesus, it is not ancient enough by a significant margin and it doesn't fit the astrological theory correctly. To put it another way, the Magi would not have been interested in such conjunctions because no one believed they had such importance until centuries later. In the same way, scientists didn't know about the importance of the Higgs boson until it was theorized in the 20th century; no one would expect Isaac Newton to incorporate such a hypothesis into his work, so we would not expect astrologers to incorporate theories before they were invented. Add in the fact that the hypothesis does not naturally lead to the 7 BCE conjunctions, and the

---

[68] Kennedy, Pingree, *The Astrological History of Masha'allah.*
[69] Kennedy, Pingree, *op. cit.* p. 72.
[70] Adair, "The Star of Christ in the Light of Astronomy", *Zygon* 47, 1 (March 2012): 7-29, esp. pp. 9-12.

chances of the Magi caring about the astrological events then are very small. What we know about the Zoroastrian religion before then is also damaging, but that will be discussed in Chapter 8.

So the actual astrological theory that would give weight to a conjunction of Jupiter and Saturn comes too late, but most Star researchers are not aware of the theory of great conjunctions anyway, and their arguments are independent from this. What is mostly stated is that Jupiter is the planet of kings, while Saturn is the planet of the Jews. Also, it is alleged that the constellation of Pisces (the two fish) was similarly associated with the Jews. On this line of reasoning, the Jupiter-Saturn conjunction stated there was a new, royal event concerning a Jewish king, supposedly his birth.

Several attempted lines of evidence are brought in to support the stellar connections, but the arguments that prove to be particularly faulty concern the constellation of Pisces. The only source provided for the link between this sign and Judaism, Judea, or the Jewish people is a late medieval rabbi, Issac Abrabanel of Spain, with some suggesting that he inspired Kepler.[71] Abrabanel writes towards the end of the 15th century, so he is literally 1500 years too late to have any likely knowledge of archaic astrological beliefs.[72] Rather, the rabbi shows no signs of holding onto an ancient tradition as he needs to argue for his association between Pisces and Judaism and his chronology to have things match up with the

---

[71] Hughes, *The Star of Bethlehem: An Astronomer's Confirmation*, pp. 68, 96, 184-186; Roy Rosenberg, "The 'Star of the Messiah' Reconsidered", *Biblica* 53, 1 (1972): 105-109. Hughes and others have spelled the rabbi's name 'Abarbanel' among other variants.

[72] Molnar, *The Star of Bethlehem: The Legacy of the Magi*, pp. 27-8 give a few arguments not to consider Abrabanel's hypotheses as fitting to classical times.

birth of Moses. His position is further undermined by other rabbis from around the same time and earlier in the High Middle Ages who all contradicted each other on this very point. Some thought the constellation of the Jews should be Taurus, others Aquarius, and some the entire Earthly Trine (Taurus, Virgo, Capricorn). This allowed another famous Jewish scholar of the early modern period, Azariah dei Rossi, to argue that using astrology is a futile pursuit (and with it, predicting the coming of the Messiah).[73] So why trust Rabbi Abrabanel any more on Pisces than Levi ben Gerson on Aquarius as the constellation of the Jews?[74] In fact, why assume they know anything about what non-Jews a millennium ago thought on occult matters? Importantly, and perhaps unknown to most Star researchers, Abrabanel and these other rabbis were using the great conjunction hypothesis themselves, so all their speculations come after its introduction into Europe in the 12th century or so. It has nothing to do with ancient beliefs but medieval concerns. There is nothing worth calling evidence that favors Pisces as the constellation representative of the Jews to eastern soothsayers.

But what about Saturn? Can any evidence be provided that this planet was linked with Judaism? Here two major sources are considered, and both are ancient. First is the prophet Amos from the Old Testament, one of the less famous Hebrew prophets but perhaps the oldest. Second is the Roman historian Tacitus from the second century CE. Let's consider them in order.

---

[73] Azahiah ben Moses dei Rossi, Joanna Weinberg, *The Light of the Eyes*, pp. 548-550.
[74] Bernard R. Goldstein and David E. Pingree, *Levi ben Gerson's Prognostication for the Conjucntion of 1345*, pp. 31, 37.

The verse in question for the Old Testament is Amos 5:26: "You shall take up Sakkuth your king, and Kaiwan [Saturn] your star-god, your images, which you made for yourselves." The verse is cited by Peter in Acts 7:43, so it was not obscure in the first centuries of Christianity. However, the context of the passage is not about Saturn representing the Jews or the planet as worshipped in Judaism. It was about the Jews failing to worship correctly; worshipping the star-god Kaiwan was sinful, not kosher practice. In fact, God specifies in the following verse how this will be a sort of apostasy and leading to exile, a punishment. Earlier in the same chapter (vv. 8-9), God is said to control the stars and the coming of day and night; he is the Lord of Hosts (v. 27, meaning he rules the sky). The point is that Kaiwan/Saturn is *not* God, who is far beyond any of the stars in the sky, and God has cursed his people to worship an idol instead. This means to use Amos 5:26 as support for Saturn being the Jews' planet is to assume precisely the reverse of what the text reads; the exact opposite conclusion ought to be reached. If one wonders why the star-god Kaiwan/Saturn was specified rather than any other planet, there may be a simple reason: in Hebrew, the word for 'star' is *kockab* (כוכב), so the author may well just be using alliteration between *kockab* and Kaiwan and nothing more.

As for Tacitus, his talk of Saturn and the Jews is related to various speculations about why they had their Sabbath on the day of Saturn (Saturday).[75] In actuality, the entire line of reasoning that Tacitus reports (and which he does not actually believe) is anachronistic. It was not until after the calendar reforms of Julius Caesar

---

[75] Tacitus, *Histories* 5.4. Cf. Dio, *Roman History* 37.16.2,4; 49.22.4; 66.7.2; Frontinus, *Stratagems* 2.1.17.

and perhaps not until his adopted son Octavian (Caesar Augustus) came to power that the weekdays were connected to the planets; the seven-day week was itself a product of Julius Caesar's calendar.[76] In the East, there were also no seven planet-named days of the week, making the speculation even more anachronistic and culturally irrelevant.[77] Even ignoring this, Tacitus does not say anything about the planet of the god Saturn representing the Jews. If anything, people were saying that Jupiter was the god of the Jews, not Saturn.[78] The entire line of Roman guesswork is very much out of place and has nothing to do with what anyone in the Middle East would have thought. Tacitus provides no evidence at all for any connection between the Jews and Saturn, at least for Eastern magi.

With these particular sources found to be wholly lacking in their utility to support the premise, a more general point ought to be observed. One should consider why anyone in the East would associate one of the planets with the Jews. There are only seven planets in the ancient system (Sun, Moon, Mercury, Venus, Mars, Jupiter, Saturn), and there are lots of potential regions, kingdoms, or civilizations that could have something represent them; the same can be said for the twelve zodiac signs. But why would we think that Babylonian or Persian astrologers would have cared so much about the small area between traditionally more powerful regions (Egypt, Phoenicia, and Syria/Assyria) and assign it a

---

[76] Barton, *Ancient Astrology*, p. 52; Roger Beck, *A Brief History of Astrology*, p. 73; Eviatar Zerubavel, *The Seven Day Circle: The History and Meaning of the Week*, pp. 15f.
[77] Zerubavel, *The Seven Day Circle*, p. 14.
[78] Augustine, *Harmony of the Gospels* 1.30. Augustine also denies that the Jews worship Saturn multiple times. Cf. *Contra Faustum* 18.2,5; 20.13; *Harmony of the Gospels* 1.42.

planet? The importance of Judaism seems to be much more in the eyes of the modern Star researcher, and that has become anachronistic when placing the same importance of this religion among Eastern sages. So perhaps it's not that telling that no Eastern record has been put forward to show that *any* star, constellation, or planet was specifically assigned to the Jews or Palestine.

This problem then impacts upon the issues in the next section about horoscopes.

Part B: Horoscopes

So far, Star of Bethlehem hypotheses have looked at a singular object or conjunction. Instead, the horoscope that astrologers have used for millennia consider the location of all the planets as well as the zodiac constellations (or 'signs'). The oldest horoscopes come from the fifth century BCE in Babylonia. These horoscopes come in ancient cuneiform writing and date up to the first century BCE.[79] However, almost all our knowledge about using horoscopes comes from Greek sources, a system that came to exist after the conquests of Alexander the Great and greater intellectual contact between East and West. It would seem that the methods of Western horoscopic astrology began in the second century BCE in Egypt and then spread from there. The oldest known Greek treatise on the subject is that of Nechepso and Petosiris (pseudonymously using the names of ancient Egyptian figures) which exists only in fragments, while the oldest extant book on how to do horoscopy comes from Marcus Manilius from around 14 CE, though the work is unfinished. Later, we have other astrologers that detail their ideas about astrology and horoscopes, notably Claudius Ptolemy of Alexandria, Egypt, his near-contemporary Vettius Valens in Antioch, Syria, and the Christian convert Firmicus Maternus.

---

[79] Rochberg, *Babylonian Horoscopes*.

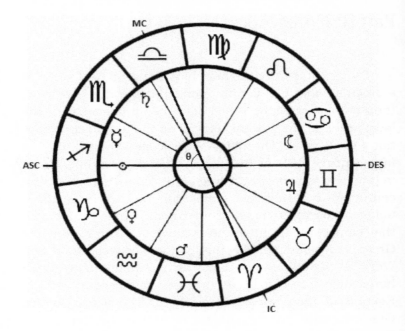

Example horoscope with cardinal points and positions of planets

Our Western sources provide many ways to look at the relative position of the planets in the sky, and as discussed in Chapter 2 it makes figuring out anything of what would have been meaningful to an ancient astrologer nearly impossible. However, the most thorough attempt at this in recent years has been by the retired astronomer Michael Molnar.[80] He uses many of the sources mentioned above and others in his reconstruction of what Jesus's horoscope would have been, along with his ideas of what in particular was "his Star

---

[80] Molnar, *The Star of Bethlehem: The Legacy of the Magi.*

in the east" (Matt 2:2). His discussion of ancient astrological practices is probably the best introduction on the subject with its technical details communicated in an understandable way; on the other hand, his treatment conceals the difficulty of interpreting a horoscope, especially how ambiguous and contradictory our textual sources are in figuring out how charts were used by astrologers.

However, Molnar is not the first to create horoscopes for Jesus. The earliest extant one that we have comes from the 15th century by a French cardinal, Pierre d'Ailly, and many others came to exist before the 19th century.[81] It has also been attempted by one Bible scholar in the early 20th century whose date for Jesus's birth was almost the same as Molnar's,[82] though the latter seems to have worked without knowing this. There are also a plethora of modern Jesus horoscopes crafted by professional astrologers.[83] As you can guess, these horoscopes do not match each other in time but are all somehow interpreted favorably. Hitherto I will focus on Molnar's version since his work is the most popular of the lot, the most thorough, and the best as an exemplar of the difficulty of the approach.

The date for the Star of Bethlehem that Molnar comes up with is April 17, 6 BCE, focusing on the

---

[81] Ornella Pompeo Faracovi, *Gli oroscopi di Cristo*.

[82] Heinrich Voigt, *Die Geschichte Jesu und die Astrologie: Eine religionsgeschichtliche und chronologische Untersuchung zu der Erzählung von den Weisen aus dem Morgenlande*.

[83] Max Tschudin, "Das Horoskop von Jesus-Christus—ein Versuch", *Astrologie Heute* 52 (Dec/Jan 1994/5): 8-11; J.-C. Weiss, "2000 Jahre Jesus Christ", *Astrologie Heute* 52 (Dec/Jan 1994/5): 12-16; Patrice Guinard, "L'ètoile de Bethléhem: Un scénario organize par des astrologues", *C.U.R.A.* (2002) at http://cura.free.fr/16christ.html; Dieter Koch, *Der Stern von Bethlehem* (see the appendices).

morning sky. He comes to this date based on not just a close conjunction of Jupiter and the Moon, but the Moon passed in front of Jupiter as seen from Earth, blocking it, which is called an occultation. This event happened in the early dawn rays of the Sun, which would connect with the dawn rising of the Star as implied in Matt 2:2 (Molnar particularly believes the text uses a technical term for a heliacal rising). Moreover, the occultation happened in the constellation of Aries, which Molnar argues was the constellation associated with Palestine. The other planets are also alleged to be in a formation that was particularly auspicious when considering Western texts on astrology.

However, the first major roadblock to considering the Jupiter-Moon occultation as propitious comes from Eastern records, the ones that ought to be more telling of what people in that region believed. Instead of such an occultation acting as a sign of the birth of a king, it was considered the opposite—the death of a king!

> When the Moon occults Jupiter (*Sagmígar*), that year a king will die (or) an eclipse of the Moon and Sun will take place. A great king will die. When Jupiter enters the midst of the Moon there will be want in Ahurrû. The king of Elam will be slain with the sword: in Subarti... (?) will revolt. When Jupiter enters the midst of the Moon, the market of the land will be low. When Jupiter goes out from behind the Moon, there will be hostility in the land.[84]

This comes from the *Enuma Anu Enlil*, the most important of the ancient Babylonian astrological texts

---

[84] S. J. Tester, *A History of Western Astrology*, p. 13. R. Campbell Thompson, *Reports of the Magicians and Astrologers of Nineveh and Babylon in the British Museum* (London 1900) II, 192, p. lxvii.

and was likely still used by horoscopic astrologers in the East.[85] We can see that the beliefs encapsulated in this approximately 3000 year old text were on the minds of the East's intelligentsia. The panic that such an occultation caused was so great that we have letters between scribes in the 7th century BCE warning of it in the future and what should be done to protect the monarch.[86] The trouble this sign wrought was so great that even having Jupiter close to the Moon could cause stress about the potential death of the king.[87]

Molnar defends his premise that occultations were auspicious to the Magi, and thus consistent with Matthew's story of an all-important stellar event, from these records by arguing that the Western practices he was reliant upon had replaced the more archaic ones used before in Mesopotamia.[88] Unfortunately, the only evidence that he provides says nothing of the sort, only that natal (that is, horoscopic) astrology was done by some in Babylonia.[89] In fact, as mentioned in Chapter 2, we can actually see from the cuneiform horoscopes right up to the year 69 BCE (well into the time when Hellenistic methods had come into existence) that they are significantly different, missing features that no Western horoscope would ever leave out. Molnar is simply stuck with the assumption that the astrology he examines has anything to do with that done by first century Zoroastrians. Not only does it seem unlikely that the Magi were

---

[85] Rochberg, *Babylonian Horoscopes*, pp. ix-16.
[86] Hermann Hunger, Simo Parpola, "Bedechungen des Planeten Jupiter durch den Mond", *Archiv für Orientforschung*, 29/30 (1983/4): 46-49; Hunger, *Astrological Reports* § 100.
[87] Ulla Koch-Westenholz, "The Astrological Commentary *Šumma Sin ina tamurtišu* Table 1" in Rika Gyselen, ed., *La Sciences des Cieux: Sages, Mages, Astrologues*, p. 154.
[88] Molnar, *The Star of Bethlehem: The Legacy of the Magi*, p. 39.
[89] Strabo, *Geography* 16.1.6.

practicing astrology in the same way as did the Greeks and Romans, given the current evidence, but in Chapter 8 I will prove that the Magi did not use such methods at all.

But even if we do stick to ancient Greek and Latin texts on the subject, there are plenty of problems for Molnar's thesis. That Aries has influence over Palestine exists in only one astrological geography, that of Ptolemy in around 150 CE, but he also uses that same constellation for what is modern-day France, Belgium, Germany, and Britain.[90] As pointed out in Chapter 2, all the other astrological geographies contradict Ptolemy, and not even a majority agrees on which of any constellation has influence over any particular region. In fact, most astrological geographies do not specify Palestine at all, making an already contradictory record ambiguous. Molnar also cites the astrologers Valens and Manilius for support, but the former is ambiguous at best and the latter is misread and in fact contradicts the premise Molnar defends;[91] there is no support to be found here that Aries was the constellation of the Jews and thus not evidence for what the Star of Bethlehem may have been.

Molnar also considers another line of evidence for Aries acting as the sign of the Jews, a set of coins from

---

[90] Ptolemy, *Tetrabiblos* 2.3 (62, 73-74).
[91] Valens, *Anthology* 1.2; Manilius, *Astronomica* 4.620-7, 797-8. The Loeb English translation of Manilius includes a map of the world with its astrological geography that shows that Aries was not the influencing sign of Israel. As for Valens, his mention of Coelê Syria may refer not to southern Syria but northern as it was named in the late second century CE after the reforms of Emperor Septimus Severus. But even assuming otherwise, Coelê Syria is not the Holy Land (it better fits modern-day Lebanon), and Phoenicia may more accurately fit Israel to Valen's geography.

the city of Antioch on the Orontes,[92] the third largest city in the Mediterranean World in the first century and capitol of Roman Syria. The coins in question come from the first half of the first century CE and on one side include Aries. Molnar conjectures that these coins were minted upon the annexation of Judea into the Province of Syria in 6/7 CE, and the coins use Aries to symbolize the recently acquired region. In Molnar's theory, the constellation was chosen because it was the well-known astrological sign of the Jews and their land, and thus Aries was the constellation the Magi would have been looking to, and so this is where to search for the Christmas Star.

However, this entire line of speculation is simply in contradiction of what the oldest coin says; the coin declares it is "of the metropolis of Antioch", mentioning nothing about Israel. Using a Jewish symbol would be very odd. It would be like the US state coins including the Statue of Liberty labeled 'Michigan' or Euro coins with an Italian minting and label celebrating France's win in the 1998 World Cup. Moreover, the earliest coins were civic coins, for city purposes, not the legate class of coins that were ordered by the provincial governor, so there is no reason to think they have anything to do with celebrating the annexation of Judea. In fact, Molnar's source for dating the earliest coin puts it the year *before* the events that caused Israel to become part of Syria.[93] Assuming Molnar's source is wrong, there is still the difficulty that these sorts of coins were minted in Antioch with Aries on the reverse for centuries after the annexa-

---

[92] Molnar, *The Star of Bethlehem: The Legacy of the Magi*, pp. 52-53.
[93] George MacDonald, "The Pseudo-Autonomous Coinage of Antioch", *The Numismatic Chronicle* 4 (1904): 110-111.

tion and even after Palestine was in a different province; they also include the city goddess, Tyche, further indicating that the coins were for and about Antioch.[94] Molnar even provides examples of coins from Antioch with the same astrological symbol during times when Judea was a client kingdom and not part of any Roman province.[95] There is simply nothing here but an erroneous use of speculation to generate data. Far more likely, Aries was representative of the city because of the horoscope of the founding of Antioch; according to city tradition, the town was founded in late April (22 Artemisius in the old Macedonian calendar) in 300 BCE around dawn,[96] and the Sun and ascendant of the horoscope would have been in Aries.

There are other arguments Molnar has for Aries being Israel's star sign,[97] but they are also speculative. For example, he claims that common knowledge of Emperor Nero's horoscope was what caused people to believe he would rule in the East and conquer Jerusalem, but this is almost impossible since emperors rarely published their horoscopes for fear they be used against them. Moreover, the theory Molnar uses is dependent still on Ptolemy's astrological geography, thus Molnar's "evidence" isn't even independent confirmation of what sign related to Israel. Molnar also relies on details from

---

[94] Examples include, from the British Museum Catalog (BMC), 4, 21, 446, 527, 535.1, 537, as well as Prieur 167.
[95] Andrew Burnett, Michael Amandry, P. P. Ripollés Alegre, and Marguerite Spoerri Butcher, *Roman Provincial Coinage* 4286-4287. On the status of Israel at this time with King Agrippa II and a Roman procurator for tax purposes, see Josephus, *Jewish Antiquities* 20.104, 137-138.
[96] Malalas, *Chronographia* 8.12; Sacha Stern, *Calendars in Antiquity: Empires, States, and Societies*, p. 243 n. 27.
[97] Molnar, *The Star of Bethlehem: The Legacy of the Magi*, pp. 104-116.

Suetonius on Nero, perhaps the worst source for this emperor—including the point that Molnar needed confirmation on.[98] Molnar also thinks that the astrologer Firmicus in the fourth century had Jesus's actual horoscope but didn't reveal this knowledge even after becoming a Christian. Unfortunately, the details of the horoscope mentioned by Firmicus don't match Molnar's horoscope for Jesus, while it best matches that of Caesar Augustus.[99] Also, considering that early Christians could not even agree on the year Jesus was born, or the day, it is baffling that there was documentation in the fourth century that would have given the correct time of birth to the hour!

This is sufficient to show that there is no evidence, besides Ptolemy's contradictory statements, to support Aries as the constellation of Judea. The same can be said for using ancient records to support any sign as that of Israel in the sky, and as stated in the previous section of this chapter it is also strange to think astrologers could care enough to give this one region such prominence in their astrological geographies; the importance of Judea is an artifact of the importance it has to modern researchers.

So far we haven't even gotten much into interpreting Molnar's horoscope for Jesus and how it had all sorts of auspicious characteristics for the person born under those stars. Sure, the occultation of Jupiter by the Moon

---

[98] K. R. Bradley, *Suetonius' Life of Nero: An Historical Commentary*, p. 247; Andrew Wallace-Hadrill, *Suetonius: The Scholar and his Caesars*, pp. 63-64; Barry Baldwin, *Suetonius*, pp. 174-180. B. H. Warmington, *Suetonius: Nero*, pp. 76-78.

[99] Firmicus, *Mathesis* 4.3.9 says that the moon is full and waxing, but Molnar's Jesus horoscope has a New, waning moon. Similarly Jupiter was not in its sign of exaltation, nor is it in trine aspect with the Sun.

was seen as terrible in the East, but what about the rest? Unfortunately, you can see just how easy it is to reinterpret the same horoscope into something terrible. For example, the Moon is approaching the Sun and is close enough for a conjunction by ancient astrologers, which is said to bring on sickness and insanity;[100] in addition, the Moon moves away from Saturn, bringing on elephantiasis.[101] Considering Mercury, because it and Saturn were in or near feminine signs (Taurus and Pisces), the male is predicted to become a eunuch or prostitute, which is exacerbated by the Moon moving away from Venus towards Mercury.[102] One can go on, using a variety of sources and potential correlations from those works to interpret a given horoscope any way that is desired, which is entirely consistent with the results from studies of modern astrologers as mentioned in Chapter 2. That makes the entire method worthless.

But what about using a horoscope in general? When considered more carefully, the process of using the horoscope to predict the birth of anyone is completely backward. A horoscope is drawn up based on the location and time of a birth, and from the chart you predict things about that child's personality and future. In other words, horoscopes are made *after* a birth, not before, and they do not predict births. So even assuming that Western astrology was practiced by the magi (it wasn't), and the Greek and Latin records were not contradicting each other (they do), and our ability to interpret them was possible and unambiguous (it is certainly not!), it wouldn't matter anyway because the entire method is being done wrong right from the

---

[100] Firmicus, *Mathesis* 4.5.1.
[101] Firmicus, *Mathesis* 4.9.6.
[102] Firmicus, *Mathesis* 3.9.1; 4.13.1-2.

starting gate. Making something more and more wrong doesn't suddenly make it right!

At this point, it may be worthwhile switching from considering how a horoscope could have been interpreted to how the theory could fit Matthew's description. There is also trouble at the start because Molnar's occultation of Jupiter was happening so close to the Sun it would have been impossible to see. While the astrologers could have calculated the event, the Gospel does state that the Magi had seen the Star. But let us instead consider the aspects of horoscopic theories rather than Molnar's particular version.

If, in fact, we can get astrology to fit the details of Matt 2:9, then it may encourage us to keep the planetary interpretation hypothesis alive. Again, the biggest proponent for this in detail is Michael Molnar who sees Matthew using actual, technical jargon from astronomers and astrologers. For example, Molnar says the word for "going before" in Matthew is the same or related to the one used for retrograde motion, *proegeseis* (προηγήσεις); similarly, "stood" is related to the word for the stationary point of a planet.[103] On Molnar's reading, Jupiter is the Star which is first in retrograde motion and then reaches its stationary point while the Magi are on the move to Bethlehem.

Unfortunately for him, his arguments about terminology show that, with all due respect, he doesn't know what he is talking about. Already in the published literature has a biblical textual critic, J. Neville Birdsall, shown that Molnar is completely wrong on all substan-

---

[103] Molnar, *The Star of Bethlehem: The Legacy of the Magi*, p. 90.

tive points when it comes to "went before" and "stood".[104] In fact, it seems Molnar cannot tell the difference between a noun and a verb in Greek, let alone details of conjugation and declension. Molnar says that *proegeseis* means "to go before", but the term is not a verb but a noun (and he provides the plural rather than the singular, *proegesis*), and even so it's not the word used by Matthew or even closely related to it (see below). There is simply no correlation between the terms used by Matthew about the Star's motion and the terms used by astrologers.

However, Molnar is not the only one to argue that the terms are about planetary motion, and one other person in particular is even worse in his understanding of classical languages. A former Jesuit astronomer at the Vatican Observatory, the Hungarian priest Gustav Teres argues similarly to Molnar, though the two seem to have concluded independently.[105] But his reasoning shows how uninformed he is. For example, the term used by Matthew for the Star "going before" the Magi, *proago*, is claimed to be a different verb, *proegeomai*, as Molnar had done, minus his noun-verb confusion. The verbs are not related, and most of their similarities come from their prefix (*pro-*) and the rules for how verbs are conjugated. The root verbs are *ago* (ἄγω) and *hegeomai* (ἡγέομαι), which are obviously different, as noted by Birdsall. Teres also claims that the verb *proegeomai* is in the aorist imperfect tense. If you are a classicist, you are probably doing a double-take with that statement. Everyone else should as well because there is no such thing; Teres has

---

[104] Owen Gingerich, Michael Hoskin, David Hughes, and J. Neville Birdsall, "Review Symposium: The Star of Bethlehem", *Journal for the History of Astronomy* 33, 4, (2002): 386-394.
[105] Gustav Teres, *The Bible and Astronomy*, p. 20.

conflated two different verb tenses. In Latin, the equivalent would be the perfect imperfect—what I call the round square—tense. If the Jesuits are teaching their students Greek, not all of them are remembering their lessons. Teres also tries to look at a Syriac translation of Matthew (which is translated from the Greek) because it is then in the same language family as Hebrew and hence closer to Matthew's original Gospel. Not only is there no solid evidence that Matthew was originally in Hebrew (his use of Mark in Greek belies the hypothesis), but the method is absurd; it's like translating the current book into Hebrew to know what it was like in the hypothetical original Arabic. This is nothing but a ramshackle intellectual edifice built upon foundations of quicksand.

So, the linguistic argument isn't working out. But even contextually there is nothing to work with. In the case of retrograde motion, which is alleged to conform to Matthew's statement that the Star "went before" the Magi, the planets move against the background sky from east to west. However, the Star is said to "go before them", the Magi, who are traveling from north to south. To use terminology from physics, Matthew has established his reference frame, and it is orthogonal to the one concerning the motion of planets. As for standing still, this is something so slow and imperceptible that it takes days to notice the motion again. The astronomer and astrologer Ptolemy says just as much: "stations cannot be fixed at an exact moment, since the local motion of the planet for several days both before and after the actual station is too small to be observable."[106] There is no way the Magi could have specified the Star had stood

---

[106] Ptolemy, *Almagest* 9.2.

still in the timeframe of their walk to Bethlehem from Jerusalem, a matter of perhaps a couple of hours. This point can't be emphasized enough, since it effectively pulls the rug from under the feet of any physical explanation. One of the few claims of Matthew with regard to the Star is clearly incoherent with respect to the laws of nature. But again, even ignoring this, the reference frame is that of the Magi, and even if a planet was at its station it would still travel with the rest of the stars in the sky from east to west, never stopping relative to the ground.

While in Chapter 7 the details of the motion of the Star will be specified, it should be clear from the above considerations that normal planetary motion has no connection with the description of Matthew. It is only by ignoring the context that it even seems plausible, but even a cursory look shows the significant observational impossibilities. I haven't yet considered what is meant by "stood over" where Jesus was in this chapter, leaving that discussion for Chapter 7.

Let me lay out, then, exactly where this leaves us. The astrological theory of the Star of Bethlehem comes up against two major problems: there is no way to actually show some particular combination of planetary positions was in fact considered auspicious to the magi, and the details of planetary motion do not fit Matthew's context. This should be sufficient for a death sentence to be administered to these sorts of speculations.

Chapter 6: What (Other) Stars do Spangle Heaven with Such Error? Last Ditch Efforts

Back in the 1950s, journalist and Bible enthusiast Werner Keller observed "[a]nything that has ever moved across the canopy of heaven, as well as much that has only existed in men's imaginations, has been dubbed the Star of Bethlehem."[107] And so, there are many other hypotheses that have been floated by various proponents that have neither found traction within the public sphere nor within the intellectual literature. However, both for the sake of completeness as well as to show that the other avenues proposed will not work better than the main ones discussed in previous chapters, I will talk about some of the odd ideas, including the outright bizarre yet non-miraculous.

One odd idea is that the Magi discovered that one of the stars in the sky was a variable star.[108] A number of the stars in the night sky change in brightness for a variety of reasons, most commonly because there is actually a star pair; two stars orbit about their center of gravity and can periodically block the light of the other, making the combination seem dimmer. One such star is Algol, situated in the constellation of Perseus where the head of Medusa is held. I have personally observed this dimming, but the dimming and return to full brightness

---

[107] Werner Keller, *The Bible as History*, p. 325.
[108] Costantino Sigismondi, "Mira Ceti and the Star of Bethlehem", *Quodlibet: Online Journal of Christian Theology and Philosophy* 4, 1 (Winter 2002).

takes several hours and may not be noticed if one isn't aware of its variations. There are three major problems with such a theory: there is no reason to think a variable star would translate into "a king is born"; no variable stars are recorded by ancient astronomers (the star catalog of Ptolemy, inherited from Hipparchus, shows no sign of such knowledge and are not unambiguously known until the Scientific Revolution); and such a star cannot move in the way Matthew describes. The variable star hypothesis is basically just as flawed as the nova hypothesis.

Another suggestion is the discovery of a planet, namely Uranus.[109] But this has the same flaws as the variable star hypothesis: no reason to think it connects to the birth of a king, Jewish or otherwise; it doesn't move like Matthew's Star; and there is no record of the planet being discovered until the advent of the telescope, such as the one used by William Hershel in 1781. Uranus can enter the lowest reaches of what is naked-eye observable, but it is still almost impossible to find unless you know where to look, something that didn't happen until it was discovered.[110] To give it the needed astrological power (and this is also the case for the variable star thesis), it is connected to the Jupiter-Saturn conjunctions in Pisces where Uranus was residing at that time. But this assumes the astrological importance of that conjunction, something found to be wanting, as shown in Chapter 5, Part A.

---

[109] George Banos, "What the star of Bethlehem Uranus", *Astronomy Quarterly* 3, 12 (1980): 165-168.
[110] If you know where to look, you can see very dim objects. I have seen Venus during the day, but I had to know almost exactly where to look.

Instead of planets, asteroids have been alleged to have been discovered. Little more needs to be said: there is no evidence that any were observed then; there is no reason to think it means "king of the Jews was born"; and the motion doesn't fit what Matthew describes.

Two more diffuse light sources have also been considered: the aurora borealis (or Northern Lights) and the zodiacal lights.[111] The former is due to high-energy particles from the Sun interacting with the magnetosphere of the Earth. Though most prominent in the arctic, they can be seen much farther south, even to states such as Texas. The zodiacal lights are due to sunlight bouncing off dust in the solar system that is concentrated in the plane of all the other planets going about the Sun;[112] it can become a diffuse pillar of light at dawn and dusk. The former light source doesn't fit well since, if anything, it will be seen in the north, so it won't rise in the east or even be seen in the south towards Bethlehem. As for the zodiacal lights, the oldest observations come from Muslims, and otherwise it was unknown before.[113] Like the aurora borealis, the lights cannot "go before" anyone, nor arrive at a place, stop and then stand over a locale. The main proponent of the zodiacal light hypothesis, the late Austrian astronomer

---

[111] Konradin Ferrari d'Occhieppo, "The Star of Bethlehem", *Quarterly Journal of the Royal Astronomical Society* 19 (1978): 517-520; "Neue Argumente zu Aufgang und Stillstand des Sterns in der Magierperikope Matthäus 2, 1-12", *Sitzungsberichte der Österreichische Akademie der Wissenschaften*, Abteilung II. Mathematische, Physikalische und Technische Wissenschaften, 206 (1997): 317-344.

[112] Interestingly, the study of the motion of this dust earned a Ph.D. in astrophysics for Brian May, the lead guitarist for the band Queen.

[113] Donald W. Olson, "Who First Saw the Zodiacal Light?" *Sky & Telescope* 77, 2 (Feb 1989): 146-148.

Koradin Ferrari d'Occhieppo, used the planet Jupiter for some things, such as the stopping, and the lights for others, namely the "going before" aspect of the Star. But this all ignores the singular "Star" in Matthew, which also doesn't fit a diffuse light cloud on the horizon.

An interesting phenomenon that remains uncertain in origin has not been properly published in relation to the Star, but astronomer David Hughes knows of some who have considered explaining the Star with ball lightning.[114] These objects are bright and spectacular, manifest themselves in our terrestrial plane, and can move about. They are also said to explode after a time. So, here is something star-like, able to move about, and it could also stand in place (potentially). However, these objects are short-lived on the order of seconds to minutes, so it couldn't "go before" the Magi for the entirety of their trip to Bethlehem. Also, it couldn't be the same object that rose up in the eastern sky, and we cannot say how it could have made anyone think "king". Furthermore, such an explosion would probably have made it into Matthew's record, if not for theological reasons but for the fact that it would be awesome, and the spectacular is something Matthew does not shy away from (zombies anyone? Matt 27:52-3). So ball lightning is probably a no-go.

Now, so far these ideas have been coming from either biblical scholars or scientists, but some theories are more on the fringe yet still try to explain the Star. One concerns the precession of the equinoxes.[115] This is an effect of the wobble of the earth much like a spinning

---

[114] David Hughes, *The Star of Bethlehem: An Astronomer's Confirmation*, pp. 170, 2f.
[115] Garry T. Stasiuk, "The Star of Bethlehem Reconsidered: A Mythological Approach." *Planetarian* 10, 1 (1980): 16–17.

top; the axis of the planet draws out a circle while remaining about 23 degrees tilted related to straight up from its orbit about the sun. This cycle lasts about 26,000 years and is credited to have been first discovered by Hipparchus of Rhodes in the second century BCE. What is noticeable to earth-based observers over a considerable period of time is that the stars slowly drift together in the same direction when compared to where the Sun is at a given day of the year, and we will eventually have a different North Star. This has the effect of changing where the Sun will be among the background stars on the important first day of spring. This drifting of where the Sun is on the first day of spring (the vernal equinox) has an effect on astrological predictions; for most modern astrologers, the signs (such as Aries, Pisces, Leo, etc.) no longer line up with the actual constellations you see in the sky because they base their zodiac on where the Sun is at the vernal equinox rather than the actual stars. This means that the Sun sign you have in the newspaper (such as Sagittarius for those born between Nov 22 and Dec 21) does not indicate where in the sky the Sun is for those dates (the Sun is more likely a Scorpio). Many skeptics have pointed this out though not knowing that astrologers have been aware of this for millennia.

But how can this possibly explain the Star? Well, the theory has to rely on a bit of questionable scholarship from the 1960s, a book called *Hamlet's Mill* which postulated a much, much earlier discovery of the precession of the equinoxes, and this knowledge pervades myths from all over the world. Since the hypothesis requires a hyper-diffusion model of culture with one super-culture from the past that left no archaeological record, and its methodology is very loose with the facts, it has no standing among scholars. Nonetheless, the idea is that Jupiter and Saturn had their conjunction where the Sun would be on the first day of spring, which is extremely important to astrologers (even though the ancient astrologers never said that), and then Jupiter is at the same important location when it becomes stationary. So, we have to rely on the stationary point idea to explain the "stood" aspect of the Star, which in Chapter 5, Part B was shown to not work (see also Chapter 7), but the thesis is also problematic because the conjunctions do not happen on the vernal equinox, just somewhat close. In fact, there was a better such conjunction in 54 CE; why no Magi then? As for how the Star "goes before them", no actual explanation is forthcoming.

And last, but certainly not least, when it comes to strange ideas about the Star is an alien UFO. Yes, this has been seriously proposed. First it was genuinely argued independently by Paul Thomas (a.k.a. Paul Misraki), Rev. Barry Downing, and R. L. Dione in the 1960s,[116] it later received support by UFO contactee George Van Tassel,[117] and it has appeared in an episode of *The History Channel's* (now on *H2*) continuing scholastic embarrassment, *Ancient Aliens* (Season 3, Episode 8). Because of its popularity, the show may make the UFO hypothesis the trendiest one for the Star, at least amongst a certain crowd.

However, not all in the alternative scholarship world buy into UFOs. For example, Douglas Elwell, who has a master's degree from Wheaton College in biblical studies,

---

[116] Paul Thomas, *Flying Saucers through the Ages*, pp. 45, 138. Barry Downing, *The Bible and Flying Saucers*, p. 134; R. L. Dione, *God Drives a Flying Saucer*, pp. 44-45, 101.
[117] George Van Tassel, *When Stars Look Down*, p. 140.

argues that the Star was the reappearance of Planet X, which itself was a product of a collision with the proto-Earth and an interstellar object in the distant past.[118] In effect, it is a meshing of the ideas of Immanuel Velikovsky's *Worlds in Collision* with New Testament fantasy. Also related is a supposed connection between the Christmas Star and Nibiru,[119] the imaginary planet of pseudo-scholar Zecharia Sitchin and his misreading of Sumerian texts. Besides the unlikely solar system history, the thesis is still not unlike the nova or Uranus hypotheses, and it thus can no more explain the Star than those failed efforts.

Still other alternative writers are more traditional, in a sense, when it comes to ideas about the Star. The New Age *Urantia Book*, for example, provides all sorts of biographical information about the life of Jesus, including his Nativity. The massive tomb states that the story of the Star is a legend but is based on the conjunction of Jupiter and Saturn.[120] In an interview on the radio program *Coast to Coast AM*, novelist and ufologist L.A. Marzulli, who otherwise believes the Bible tells of inter-dimensional beings, also argued for a conjunction as the Star of Bethlehem rather than a flying craft, but he went for a conjunction between Jupiter and

---

[118] Douglas A. Elwell, *Planet X, the Sign of the Son of Man, and the End of the Age*.
[119] Barbara Hand Clow, *Chiron: Rainbow Bridge between the Inner and Outer Planets*, p. 21; Al McDowell, *Uncommon Knowledge: New Science on Gravity, Light, the Origin of Life, and the Mind of Man*, p. 185.
[120] *Urantia Book*, paper 122.8.6-7. About this book, including its take on the Christmas Star, see Martin Gardiner, *Urantia: The Great Cult Mystery*, esp. pp. 206-209.

the star Regulus in 3 BCE.[121] So, you can't please everyone it seems.

But, strange as this idea is, at the very least the UFO can match up with what Matthew says. A ship can appear like a star when far away and assuming it is reflecting or sending out light. If the ship wanted to, it could be seen in the eastern skies. And a ship can travel from north to south, stop in place, and hover over a location, even a very specific one—at least according to Spielberg. If aliens were doing this, perhaps then they could also have had chats with the Magi (or the aliens used "thought transference" as Van Tassel believed) so the Magi could know what they were up to, why it's important to follow, and so on. There is the issue though of how the Magi would have seen the UFO over a house and still not notice it is a ship, so the fit to Matthew's Gospel is still not perfect. Of course, the real sticking point is that we don't know if advanced civilizations even exist in our galaxy besides our own, and the immense distances between stars make travel prohibitive without the laws of physics being curiously out-maneuvered by unknown technologies. The issue of motivation for the aliens is also something not to be put aside. But the only way to properly discount the UFO hypothesis is either to prefer the miraculous version of the story or deny the historicity of the event altogether. So, if you buy into the alien hypothesis, you will need to consider Chapter 8.

---

[121] A summary of the show with Marzulli can be found at http://www.coasttocoastam.com/show/2010/12/22.

PART III: FATAL PROBLEMS

Chapter 7: Failure of all Natural Hypotheses

While the individual chapters about particular ideas regarding the Star can show the significant weaknesses of those hypotheses, this chapter can, on its own, show why none of them will ever be able to work. It is simply a problem of the details of Matthew's account; and many of these issues various Star researchers try to reinterpret while doing harm to the context and meaning of the words. Part of the problem is that most recent authors on the subject are unable to read the story in its original language, Koine (common) Greek. While many do talk about particular words or phrases from the manuscripts, the understanding is not broad, and they are still primarily reliant on translations. As such, I will need to talk about the words and phrases carefully to make the point of what the text does indeed say.

The phrase that perhaps receives the most attention concerns the Star "at the rising" (*en te anatole*), which has been variously interpreted to mean rising at sunrise, rising at sunset, or a very particular rising with the sun called a *heliacal* rising—the first visible rising on the eastern horizon before being blotted out by the sun. The first person to argue that Matthew was actually using a sophisticated phrase concerning the heliacal rising was the biblical scholar Heinrich Voigt in 1911.[122] However, this was contested by one of the foremost experts on ancient astrology of the time, Franz Boll, who stated that the argument was "naïve".[123] Modern translators of ancient Greek astrology texts also read it the same way, that it is not a special phrase from astronomers or astrologers in antiquity.[124] But this matters little in the scheme of this chapter, because there are plenty of natural phenomena that can conform to this phrase in any interpretation found in translations or Star of Bethlehem literature.

It is when considering verse 9 of chapter 2 of Matthew's Gospel that things get hairy. The word that describes how the Star "went before" the Magi is the verb *proago*, which means to lead forward. But in the context of Matthew, it is even more specific because the verb takes a direction object—that which the verb is acting on—and that direct object is clearly the Magi. As such, the Star was leading the Magi, bringing them forth to their destination; the Star is doing more than standing in a certain direction or even moving about, but it is actually leading the Magi on. This can be similarly seen in Matt 21:9 with the crowds that "went before" Jesus when he entered into Jerusalem; it was like a parade procession. This point is expressed in some translations of the Bible better than in others, such as the Holman Christian Standard, the New Living Translation, The

---

[122] Heinrich Voigt, *Die Geschichte Jesu und die Astrologie*.
[123] Franz Boll, "Der Stern der Weisen", *Zeischrift für die neutestamentlich Wissenschaft*, 18, Jahrg., Heft 1/2 (1917): 40-8.
[124] Courtney Roberts, *The Star of the Magi: The Mystery that Heralded the Coming of Christ*, pp. 120-1.

Message, as well as translations by particular scholars such as Hugh Schonfield and Robert Miller.[125]

But even ignoring the particulars of the meaning of the word, the context has the Star "going before" the Magi who are traveling from north to south, while all objects in the sky travel instead from east to west. Perhaps one can imagine a scenario with the road to Bethlehem turning in just the right way as the Magi walk at just the right speed with an object in the sky in just the right position so it remains over Bethlehem over the time journey. But even so, that does not fit the motion specified by the verb *proago*.

And indeed it does move, and then it stops. This is fixed by Matthew saying that the Star was "going before" the Magi "until" something happened. That "until" (using the Greek word *heos*) means there is a termination of the previous circumstance. If in many readings of Matt 2:9 by scientists that the Star is simply to the south and so in the direction of Bethlehem, then Matthew is saying that situation stopped when the Star "stood over where the child was". This makes no sense on the natural phenomenon hypothesis. This "until" was noted by David Strauss in 1835 when he challenged many sorts of rationalizations of miracles from the Bible in his day.[126] The only person who has tried to deal with this is the Jesuit astronomer Gustav Teres who was mentioned in Chapter 5, Part B, and who was found to be linguistically incompetent. Here it is no different.[127] He instead wants to translate *heos* to mean 'while', which is possible depending on context. Here it doesn't make sense. If we

---

[125] Hugh J. Schonfield, *The Original New Testament*, p. 58; Robert J. Miller, *The Complete Gospels: Annotated Scholars Version*, p. 61.
[126] David Strauss, *Das Leben Jesu kritisch bearbeitet*, pp. 220-53.
[127] Teres, *The Bible and Astronomy*, p. 21.

use his translation, then the text says the Star was going while it stood; it goes while it is not going. Not even a spiritualized reading by the Gnostics is going to fix this contradiction, and I am not even considering issues of grammar here, just basic logic.

Another word of interest concerns how the Star "arrived" or "came". Some translations like the New International Version do not include this because in English it seems to only add verbiage, but for our purposes it is worth noting. The operative word here is a very common verb, *erchomai*, which means 'to come/go', depending on context. As the verb is here conjugated (as a participle), it means that in an instant (the aorist tense) it came to its destination. Again, we have a verb used that describes motion. That is in utter contradiction to any hypothesis that the Star was simply seen in the south. Both *proago* and *erchomai* both clearly indicate that the Star was moving during the Jerusalem-to-Bethlehem journey. Moreover, the Star arrived at a location, one that is specified, over which it stood still.

This is now leading up to the standing of the Star, which already the word *heos* ('until') indicated was coming. The term here is *histemi*, another very common verb that means 'to stand/make stand'. The verb is again conjugated like *erchomai* to indicate that the Star came to a standstill in an instant using the aorist tense. That would disallow the stationary point of a planet which for an ancient observer was unperceivable in how long it took to stop and go again. But more importantly, there is nothing in the sky that comes close to traveling south for a couple of hours and then coming to a halt.

But perhaps it is the last part of the description of the Star that dooms any theory of an astronomical or astrological phenomenon. The phrase used by Matthew says the Star "stood over where the child was". The word 'over' here is using the preposition *epano*, but there is another aspect that is invisible when translated into English. The term *epano* has a particular meaning when it is followed by a noun, participle, or adjective in a particular declination called the genitive. Prepositions in Greek usually only take certain declinations as their partners in meaning, though some take multiple possibilities which allow for different meanings. Here *epano* when followed by the genitive takes a meaning close to what we get with one of *epano*'s root words, *epi*. This preposition is known even in English in words such as epidermis (top layer of skin) and the epicycles discussed earlier, and it has the meaning of being on top of something. And that is the sort of meaning Matthew's text implies; the Star is not simply above Jesus, it is hovering right over the house he is in. For comparison, the same construction is used by Matthew in 5:14 concerning a city *on* a hill, in 23:18, 20 about gifts placed *on* alters, in 23:22 about God sitting *on* a throne, in 21:7 when Jesus rides *on* a colt and ass which walked *on* a robe-covered road, in 27:37 about the sign on the cross *above* the crucified Jesus, and in 28:2 about the angel that sits *upon* the stone door that had closed the empty tomb. In all these cases, Matthew uses *epano* with the genitive to mean something that is on top of or slightly above an object. This meaning is consistent in other places used in the Bible, including Luke 4:39, 10:19, 11:44, and Revelation 6:8, and the same can be said for other Greek literature.

Compare this to another possible preposition Matthew could have used, *huper*. As discussed in Chapter 3, this preposition was used to describe comets that hung over cities. Josephus had done this in reference to a star over (*huper*) Jerusalem, but this can mean the object was up in the heavens. The same is done by astronomers like Ptolemy that speak of astronomical or astrological events "above the earth" (*huper gen*, ὑπὲρ γῆν).[128] Not true for *epano* with the genitive which implies close proximity. The term is distinct, and there is no astronomical or astrological work that uses this construction when it comes to objects in the sky relative to objects on the ground. If you want to check this, you can use the database program *Thesaurus Linguae Graecae* and examine all the astronomical and astrological works in Greek in the catalog; you will find that *epano* itself is very rarely used at all, and even more rarely followed by the genitive. More often the word is used with respect to the text of their volume (i.e. "as was shown *above*"), and Ptolemy used it with respect to stars relative to other stars in the heavens, and those stars were still close to each other on the celestial sphere of the night sky.[129] For example, Ptolemy spoke of the star 18(e) Ursa Major as "above the right knee" of the bear,[130] about 3.5° away. So in one of the very few places where an astronomer uses *epano* with respect to the stars, it is in relation to another nearby star in the night sky.

This demonstrates beyond reasonable doubt that Matthew is describing the Star as in close proximity to where Jesus was, which is said, in verse 11, to be a house. In another infancy gospel attributed to Jesus's brother James, the situation is also made stark as the Star enters into a cave where Jesus is and stands over his head.[131] Similarly, this is how it has been understood

---

[128] Ptolemy, *Tetrabiblos* 2.5 (76).
[129] Toomer, *Almagest*, pp. 15-6.
[130] Toomer, *Almagest*, p. 342.
[131] *Protoevangelium of James* 21.

by generations of readers, from theologians to astronomers. Let me give just two examples, both of which show that it was not due to lack of knowledge that led to the miraculous reading. First, consider St. Augustine of Hippo. Before he was a Christian he belonged to a religion that was heavily involved in astrology, and Augustine also read about astronomy and astrology and conversed with astronomers on their craft.[132] So when he said that, unlike the stars in astrology, the Star of Bethlehem left the course of the rest of the sky and came down to a newborn, and astronomy and astrology have nothing to explain,[133] we should take seriously that he understood the science involved as well as the meaning of Matthew's account.

But also consider the person so many scientists refer to when they make their own journey into Star research, Johannes Kepler. His work on connecting Jesus's birth with the great conjunction in 7 BCE certainly involved astrology, but Kepler, who in his schooling learned Greek and Latin, also specified the miraculous in the tale. "This star was not one of the ordinary comets or new stars, but by a special miracle moved in the lower regions of the air."[134] Kepler was no deist, and when he read of a miracle in his holy book, he believed it was a miracle, and he believed the Star to have traveled down low to the ground for the Magi to see and follow.

---

[132] Augustine, *Confessions* 4.1-3; 5.3. See Leo Ferrari, "Astronomy and Augustine's Break with the Manichees", *Revue des Études Augustiniennes* 19 (1973): 263–276.
[133] Augustine, *Reply to Faustus* 2.6f.
[134] Christian Frisch, Joannis *Kepleri Astronomi Opera Omnia* IV, p. 346. Translation of the following Latin: *Stella haec non fuit e numero communium cometarum aut novorum siderum, sed accessit illi privatim miraculum motus in inferiori regionis aeris.*

Countless others from antiquity, medieval times, and the early modern period can be cited, including Thomas Aquinas, Martin Luther, and John Calvin. To find these three agreeing on anything ought to indicate what the text is actually saying. I document this interpretation through the centuries and by modern scholars in a full article on the subject which includes the first attempts to understand the Star as something natural beginning in the early 19th century.[135] Today, one is hard-pressed to find any recognized Bible scholar publishing in peer-reviewed literature taking the view of the Star as something non-miraculous, the last example I can find coming from the mid-20th century. Among those that can read the text critically, there is no question about its presentation.

What this demonstrates overall is that nothing fits with what Matthew actually talks about, something few Star researchers are able to even access because classical languages are being taught less and less, especially to scientists. But Matthew talks about a Star that travels south towards a particular destination, leading on eastern sages, until it comes to its destination, stops and hangs over a particular hovel in the small town of Bethlehem. No object in the sky can do such a thing, not by a long shot. The only two natural phenomena that we could even talk about fitting such motions are ball lightning (but see chapter 6) and a UFO. However, we really are left with the impression by the author that something miraculous has happened.

At this point it ought to be clear: we have a miracle as described. So, if we accept that this is what is actually

---

[135] Aaron Adair, "The Star of Christ in the Light of Astronomy", *Zygon* 47, 1 (March 2012): 7-29.

recorded by Matthew, can we say if the events happened or not? This idea is discussed in the next chapter.

However, considering the consistency of interpretation of the miraculous nature of the Star for centuries and by modern biblical scholars, it raises the question as to why anyone thought it could be something naturally explained. I explore this in my article mentioned earlier, but here is a summary of the history of such explanations.

Up until about 1800 everyone who said something about the nature of the Star agreed on its supernatural qualities, and no exception exists until the 19th century. Why the change? For one, the premises of the Enlightenment had meant for many that miracles simply do not occur. Either they are impossible, as argued by the philosopher Spinoza, or reports of them are almost never believable, as argued by David Hume. English deists and other skeptics had an influence on many in continental Europe, and the most important example of this influence is a German scholar of the Bible, Hermann Samuel Reimarus. In his writings that were published posthumously in the 1770s, he took those beliefs about the miraculous to their logical conclusion and made a huge swath of the Bible, Old Testament and New, to be credulous clap-trap. He even suggested that the Disciples were conmen of sorts. It was he who first seems to have argued in print that the Star of Bethlehem was a fiction.[136] The published fragments of his work caused outrage in Hamburg, the religiously conservative town that saw to their first publication, and it sent shockwaves through the rest of academia.

---

[136] Gerhard Alexander, *Apologie; oder Schutzschrift für die vernünftigen Verehrer Gottes*, Vol. 2, p. 536.

In reaction to the wave of supernatural skepticism among those who still considered the Bible, especially the Gospels, to be trustworthy accounts of what really happened, an unusual compromise of ideas took place. The Bible was true, but so were the laws of nature as described by Newton and other scientists. To reconcile, the miracles of Scripture were not actual aberrations of God's perfect laws of the universe, but events which the 'noble savages' saw and misunderstood. The prime example of one of these scholars from the early 19th century was Heinrich Paulus. He famously argued that the miracle of Jesus walking on water was nothing so exciting; it was dark and foggy, so the Disciples on their ship thought they were out at sea rather than at the shore, so when Jesus was walking by on the beach they thought their master was doing the impossible. Now, imagine this sort of thinking for every miracle in the Bible, and you have the Rationalist movement among Protestants. And the logic is simple: miracles are impossible, but miracles misunderstood are possible, and if it's possible and we know in our hearts that the stories are true then what possibly happened probably happened.

Now, this movement did not last for a long time, though echoes of it still exist—if you want to see this sort of thinking in action today, look at the Ancient Alien Theory (AAT) crowd.[137] It fell out of the mainstream scholarly world by the middle of the 19th century, but it seems that this sort of thought process is going on when it comes to the Star today for many scientists. Because the idea is ultimately a Christian apologetic, a defense of the faith, the naturalistic Star exists for that purpose.

---

[137] Cf. Robert M. Price, *Night of the Living Savior*, pp. 44-45.

Some people, such as Robert Newman, are explicit that they use the naturalistic theory for apologetic ends, even when they believe in the supernatural.[138] Besides the various scientists that publish on the subject, the theory is touted around in other Christian apologetic outlets. In the late 19th century, Francis William Upham used Star research to fight atheism,[139] and those who attack the "New Atheists" do much the same.[140] Fox News anchor Bill O'Reilly has also recently used the Star as historical proof for the stories in the Gospels by arguing for a comet,[141] though his argument is even more confused on the details than in other comet literature. One of the most popular productions out there defending the physically-explicable Star comes from a team of evangelicals. Lawyer and preacher Fredrick Larson has a video presentation and website defending the thesis of Ernest Martin about the Star, which was produced by the same person who produced Mel Gibson's *The Passion of the Christ*. Larson himself was also a student of the late Francis Schaeffer, one of the most influential evangelical apologists of the 20th century. Perhaps you can see why his presentation on the Star has been broadcast multiple times on the Christian Broadcasting Network (CBN) and featured on the Trinity Broadcasting

---

[138] Robert Newman, "The Star of Bethlehem: A Natural-Supernatural Hybrid?" *Interdisciplinary Biblical Research Institute* 2001, pp. 1–16.
[139] Francis William Upham, *Star of our Lord: or, Christ Jesus, King of all worlds, both of time or space. With thoughts on inspiration, and the astronomic doubt as to Christianity*, p. 306.
[140] Michael Poole, *The 'New' Atheism: 10 Arguments that Don't Hold Water*, p. 43. Rev. Downing in his sermons and in our email exchanges also used UFO explanations for biblical miracles to combat liberal theologians and atheists.
[141] Bill O'Reilly and Martin Dugard, *Killing Jesus: A History*, chapter 1, n. 6.

Network (TBN). So we see a strong contingent of Christian apologists using the Star explained by science to defend the faith.

Now, the motivations of the scientists who promote these ideas may not be so evangelical, but it is certainly so for some. This is definitely the case for Frank Tipler whose book on the subject is all about proving Christianity through physics.[142] The evangelical Francis Collins, former head of the Human Genome Project, current director of the National Institute of Health in the United States, and a major voice that tries to find a place for Christian belief in science, cites work on the Star of Bethlehem as proof of reconciliation between science and faith, not to mention support for the historicity of the stories of the Bible.[143] On the other hand, the motivations of other scientists are probably more along the lines of their own personal beliefs and motivated reasons to support them. However, all of these explorations into the Star as a non-miraculous object stand out because the same procedure does not exist for most other miracles in the Bible, such as the Virgin Birth or the Resurrection. These Star hypotheses are a hold-over from early 19th century Rationalism, but incompletely. But why is there no effort to make the resurrection of Jesus a natural phenomenon?

What this Star research ultimately attempts to do is prove that something from the Gospels actually happened, some guarantee to modern Christians that something about the story of Jesus can be demonstrably shown to be true with the tools we use to prove galaxies are 'island universes' millions of lightyears away and

---

[142] Frank Tipler, *The Physics of Christianity*.
[143] Karl Giberson and Francis Collins, *The Language of Science and Faith: Straight Answers to Genuine Questions*, p. 108.

planets exist around other stars. And if one thing can be proven true in the Gospels, it makes all the other stories more plausibly true, including the miraculous parts, and it provides salvation for the theological message of the Bible.

Consider why this research only rarely happens outside of Christian circles and without the positive results. It is telling that no skeptics or atheists have use for the research on the Star of Bethlehem. One of the few nonbelievers to study Star research was science fiction writer and prolific author Issac Asimov. He looked at the various hypotheses about the Star as interesting what-if scenarios, but he ultimately concluded the story was a fiction.[144] The skeptical "Bad Astronomer" Phil Plait also figures the tale is likely a fiction.[145] Historians of ancient astronomy also ignore the attempts to find the Star; as the influential historian of the exact sciences Otto Neugebauer said of tales like the Star of Bethlehem, "I think it better simply to discard such sources for the reconstruction of historical data."[146] Overall, the investigations into the Star are only used to make the biblical story more plausible and more believable rather than being "explained away" like the psychic abilities of Uri Geller—the difference being that the skeptics think there is nothing that happened to be explained. So it is understandable why the story in Matthew is breached from its context and made into something that could potentially find coherence with the sciences.

---

[144] Issac Asimov, *The Planet that Wasn't*, pp. 183-194.
[145] Phil Plait, "Starry, Starry Night," in Adriane Smith, ed., *There's Probably No God: The Atheist's Guide to Christmas*, pp. 59-67.
[146] Otto Neugebauer, *A History of Ancient Mathematical Astronomy*, p. 608.

One anecdote to consider is the email conversation I have had with David Hughes, whose book on the Star of Bethlehem is still probably one of the best starting points on the subject (though I disagree with his conclusions). He basically says that it is worrisome if Matthew's account of the birth of Jesus is not historically true, even though the Evangelist seems like such an honest guy; then it throws much else into doubt. What do we believe and disbelieve, then; and are we prone to special pleading and other methodological biases in so doing; what criteria can differentiate truth from falsity? These are not easy questions to answer, but they are the questions of the historian, and we may have to leave behind a lot of stories as fiction when we do history. Obviously for someone that considers the Bible authoritative, theologically or otherwise, it is troublesome to believe that some, many, or perhaps all the stories are not factual, and that makes for some difficult introspection to understand what one should believe. This is highlighted for scientists who study the world as acting without miracles, yet they are reading about what all their studies indicate is impossible. Such anxieties explain the motivation for making sense of tales for those who already believe on faith yet must square with their reasoning and understanding of natural law.[147]

In a strange way, this situation is best attested in the writings of those that propose a UFO as the Star. It is especially helpful to have the view articulated by Barry Downing who has a doctorate in theology as well as a science background. In his book, *The Bible and Flying*

---

[147] Hughes has expressed more skepticism recently towards the story of the Star and Nativity in general, but he seems uncertain where to stand. He also likes to give talks on the subject, a perennial favorite among holiday audiences.

Saucers, he looks at the 'demythologizing' program of theologians such as Rudolph Bultmann and John A. T. Robinson. In this paradigm, the worldview of the writers of the Gospels is not that of modern cultures, especially when it comes to symbolic narrative compared to literal histories of events. Unlike ancient people, moderns do not believe in supernatural powers and principalities, and the antique believer was not scientific and critical. Bultmann, in particular, wanted to no longer believe in the mythological and instead wanted to get to the core of the meaning of the tales. There were still core historical facts, such as the crucifixion of Jesus, but faith was more about existential reflections upon the tales of Jesus rather than recounting the history of Jesus. Bultmann himself was following in a tradition of other 19th century theologians, such as Friedrich Schleiermacher, Ludwig Feuerbach, and the young David Strauss, though Bultmann also combined his approach with existential philosophy. Faith is not about believing that a series of impossible events happened, but it is about accepting the meaning and purpose of the narrative.

Downing, however, says this leaves the stories of the Bible without power. In the preface to the 1973 edition of his widely-sold paperback as well as in his sermons and our email conversations, he notes how the aliens, as emissaries of God, provided humanity with morality and spiritual purpose. He needs the Bible to be true for there to be spiritual and moral potency, not to mention the promise of eternal life in heaven, and it is the belief that we cannot trust the miraculous tales in the Bible that undercuts this. UFOs save the phenomenon of the text, then, and with it our souls. But why is believing in miracles not sufficient? As another UFO proponent puts it, then any ghost story is just as well justified[148] (not to mention other, non-Christian religions). Science (or more science fiction) is what makes the stories become realistic to a modern mind and not mere myth; there is no inconsistency in believing in light bulbs as well as supreme beings.

This can probably account for many of the efforts, past and present, to explain the Star consistently with physics and why Star research has a popular following, even all the way back in the 1930s with the earliest planetariums and their Christmas shows. There are people who want science, the best way to know things about the world, to support their beliefs, especially when science is perceived to undercut so many other religious dogmas.

Perhaps, then, it is understandable why there are these attempts to read Matthew's account about something astral, something science has a lot to talk about, as a natural light rather than a supernatural one. But as shown, one cannot analyze Matthew's account seriously and critically and still believe that there is something physically explicable behind it. Yet, even if we suppose against all else that the tale is about a real object, scientifically-explicable or otherwise, did it really happen? The final chapter will explore just how plausible the story actually is.

---

[148] Dione, *God Drives a Flying Saucer*, p. 79.

Chapter 8: Historical Issues

Consider now that the Star of Bethlehem, as told to us by the Gospel of Matthew, is a miraculous object, moving about with the aid of forces not explained by physics. Can we believe it happened? Should we? Some adherents to the truth claims of Christianity will complain that discounting a story because it is miraculous is a bias; yet, if we are to use that term, it is an appropriate bias. We tend not to believe in the miracle claims of other religions, since otherwise we would end up converting. Followers of the Abrahamic faiths do not believe in the miracle statues of India or the healing and resurrecting powers of Asclepius. Modern Christians don't believe a lot of the stories told from the Middle Ages, such as St. Christopher being transformed from a dog-head into a full human, and they don't believe in the ability of Rev. Jim Jones to walk on water and raise the dead as he was claimed to do. Ask yourself, how much evidence would you need to believe Jim Jones raised the dead? Just two congregation members' testimonies? You are probably not going to believe in him based simply on that, so why accept similarly miraculous claims from even more dubious reports? Perhaps you don't care for this history, but even for Jesus there are things said of him that you probably don't believe. Most Christians don't accept the stories of Jesus as a boy in the non-canonical infancy gospels, such as the *Infancy Gospel of Thomas* that had Jesus forming living birds out of clay and resurrecting his friends among other miracles. You probably don't believe the story in *The Gospel of Pseudo-Matthew* that has Jesus taming dragons while still an infant at his mother's breast. So if not believing in a miracle claim without evidence is a "bias", it is one shared by everyone except the most credulous. This is because we have realized that not everything we read or hear is true, and most of the amazing claims are probably false. In reality, we are all skeptical of such claims except those we already believe, but that is not a bias but a hypocrisy.

But even if we ignore what happens to the prior probability of a story that contains miraculous elements, could the story still be true? Let's even suppose for the sake of argument that some natural phenomenon can account for the Star or Matthew's account is so confused to render it useless in determining if he meant something miraculous (in which case, why believe him at all?), so we don't actually disqualify the story for having too heavy a divine hand in the action. And let's even suppose that Matthew is actually trying to tell us what really happened rather than inventing myths. Given such conditions, are there reasons to reject the story as it is?

As discussed briefly in Chapter 2, even if Matthew is actually trying to write a history or biography, we are still unsure about his sources and methods. If he is relying on people at the end of the first century (or later perhaps), and if they don't have a connection to the events, then it doesn't matter what Matthew's intentions were. As was also mentioned, the number of living witnesses for Matthew to contact by the time he wrote was zero; Joseph and Mary would have been dead due to age (Mary must have been at least 12 when Jesus was born c. 6 BCE, so she would have to be at least over 100 by the time Matthew wrote, a near impossibility in antiquity with an average lifespan not better than 50 for someone who reached adulthood; Joseph was even older and traditionally understood to be dead before Jesus started his ministry). The brothers and sisters of Jesus were also likely dead at this point due simply to age (and James is said to have been martyred before the end of the Jewish War). Considering also how Matthew relies heavily on Mark who was not an eyewitness even traditionally, let alone according to critical scholars, it doesn't look like Matthew had access to such witnesses.

Combine that with how Matthew doesn't do anything to show he was treating his sources with a hint of critical thought. After all, he says nothing of the plausibility of his story, and since it is significantly different to Luke's (contradictions or no) he ought to be saying why he trusts his sources over others. Again, that assumes Matthew is even trying to write history, but even if he was, he proves to be poor on method. Since the author of this Gospel is using even worse methodology that one of the most notorious biographers from the Roman Era, Suetonius, and Matthew's access to reliable sources is even less likely than was the case for Suetonius, we have to trust Matthew much less. So, at best, if Matthew is trying to be a biographer, his story of Jesus's birth has to be treated with significantly more skepticism than with most any other biography from antiquity—and we have biographies of people who never even existed, like Romulus.

However, let's leave even that aside and consider if the story is plausible, even if we have no qualms with the miraculous. If we bend over backwards to accommodate the claims to this degree, can we still believe the tale? There are several features that I will highlight which show, even if miracles are possible, that the story goes against what we know about the past.

Let us first consider the Magi themselves. According to Matthew, they came looking for the King of the Jews because they saw a Star in the east/as it rose and came to worship. First, this is strange because the Magi are not Jews but Zoroastrians from Persia. The magi were in fact the caste of priests in this religion that is older than Judaism, and they had their own king and their own belief in a savior figure much like the Jewish messiah. So we have to wonder why they would have given two hoots about the birth of a Jewish king which they may not have even connected to a savior figure—it is Herod in the story that connects the King of the Jews to the Messiah. Historical circumstance already makes the tale suspect, because now we need magi that want to become Jews because of a Star.

But what about Zoroastrian interest in the stars and astrology? Here we find another significant problem: before the Sassanid period in Persia (3rd-7th century CE), the Zoroastrians did not practice astrology.[149] The scriptures of the magi, such as the Avesta literature, show only the most basic of stellar interests, and it says nothing of the planets. When in the third century the astrological treatises of the Greeks and Indians were brought to Persia for translation into their language,[150] we see a visceral reaction to astrology by the priesthood. They considered all the planets, including the Sun and

---

[149] Gerard Mussies, "Some Astrological Presuppositions of Matthew 2: Oriental, Classical and Rabbinical Parallels", in Peter Willem van der Horst, ed., *Aspects of Religious Contact and Conflict in the Ancient World*, pp. 25-44, esp. pp. 28-37; Albert de Jong, *Traditions of the Magi: Zoroastrianism in Greek and Latin Literature*, pp. 397-398; R. C. Zaehner, *Zurvan: A Zoroastrian Dilemma*, pp. 147-165, 369, 377f, 400f, 410-411; *The Dawn and Twilight of Zoroastrianism*, p. 238.
[150] Bayard Dodge, *The Fihrist of al-Nadim: A Tenth-Century Survey of Muslim Culture*, vol. 2, pp. 573-575.

Moon when eclipsed, as corrupted by Ahriman, the Zoroastrian version of Satan. This was because of the irregular motion and changes of the planets, unlike the never-changing stars. Change in the heavens was a sign that those objects were corrupted, possessed by evil, and no good was to come of them. This was also, in part, why the Zoroastrians thought comets were evil (as seen in Chapter 3)—their appearance was irregular. The magi then were like many fundamentalist pastors with respect to astrology: either indifferent or hostile. This discussion lasted for some time, though later the magi came to accept and practice astrology, but only after its re-introduction into the Middle East. No direct evidence of Zoroastrian interest in astrology exists before then.[151] So, if there was some astrological portent in the first century, the magi wouldn't have noticed it; if there was a moving Star, they would have called it evil, not something to lead them to worship.

So why are there magi in Matthew's story if they had no astrological interests? There was actually a misconception about the founder of the religion of the magi. Known as Zarathustra in the Persian language, in Greek and Latin he was called Zoroaster. The latter has the Greek word for 'star' in it, so we can see why some would associate the practice with the man with *aster* in his name. There was also a blending of the differences between the magi and Babylonians, the latter who invented astrology and transmitted it to the Greeks. This is not unlike modern Westerns not knowing the difference between Arabs and Persians. In ancient Western sources, Zoroaster is called a Babylonian and

---

[151] Ehsan Yarshater, *Encyclopaedia Iranica* Vol. 2, pp. 258-259; Mussies, "Some Astrological Presuppositions of Matthew 2", p. 30 n. 31.

the founder of the magi. Because of such confusions, a vast body of literature on astrology was pseudonymously penned by Zoroaster.[152] As such, if Matthew had a source, that person seems to be someone influenced by Greco-Roman misconceptions and not relating history to us.

But suppose, nonetheless, that the Star was even more miraculous than Matthew describes. In other infancy stories, including ones about the Magi, all sorts of crazy things were happening in the East that told the Magi of the newborn king. In the recently translated *Revelation of the Magi* from the late second or third century, the Star is baby Jesus himself, who comes to spread the news of his Incarnation—talk about an early self-promoter![153] So let's suppose even more miracles, which are killing the probability of this story even if miracles happen. If all sorts of miraculous machinations were happening in Persia to get the magi interested and cause at least a few of them to travel to Palestine, then one would think that the great king they sought would have had great reverence in their land and his followers respected. Even if they did not see Jesus as the ultimate savior they could have still revered him in the way Muslims do today. Reality, on the other hand, couldn't be further removed from this. In inscriptions from the third century, we have the leader of the magi conducting persecutions of Jews *and Christians*![154] We find nothing

---

[152] Roger Beck, "Thus Spake Not Zarathustra: Zoroastrian Pseudopigrapha of the Greco-Roman World" in Mary Boyce (ed.), *A History of Zoroastrianism*, Vol. 3, pp. 491-565.
[153] Brent Christopher Landau, "The Sages and the Star-Child: An Introduction to the *Revelation of the Magi*, an Ancient Christian Apocryphon", Th.D. Thesis, 2008.
[154] Solomon Nigosian, *The Zoroastrian Faith: Tradition and Modern Research*, p. 34; Mary Boyce, *Textual Sources for the Study of*

that shows any deference for the Christian faith or the religion it flowered from, but we see figures harassing those that follow the man the Magi were supposed to have worshipped. So it seems when it comes to Magi honoring Jesus, no one told the actual magi.

The above considerations are already forcing the probability of this story down to zero with great speed, but let us suppose even against what has been noted that some group of magi were interested in some Star, that they did come and praise Jesus at his birth, and so on. Let's just say that our record of Persia isn't good enough to have on hand the actual report of the Magi's journey and things went awry in the time after their visit so that the importance of Jesus wasn't recognized (could it be *Satan!*?). But what about the consequences of such a journey to Israel, had it happened? Would this have been of note and reported and exist in our extant records of the past?

In actuality, it should and would have been noteworthy. Sure, a small group of eastern sages traveling about may not seem like much, but consider that the Magi are not just your local pastor or assistant rabbi. These people were part of a religious caste that not only had significant authority on matters of the divine, but they also were part of the selection of the king of Persia. For example, Strabo, a writer from just before the birth of Jesus, wrote that the council of magi would pick the king from familial claimants.[155] Further in the past, the father of history, Herodotus, said the magi had led an overthrow of the government with their own chosen king

---

Zoroastrianism, pp. 112-113; Philippe Gignoux, "L'Inscription de Kartir à Sar Mashad", *Journal Asiatique* 256 (1968): 387-418.
[155] Strabo, *Geography* 11.9.3.

back in the sixth century;[156] even though they failed, the magi were still given patronage by Darius, the king who had defeated them.[157] Other authors from the past considered the magi a little less than kings.[158] Perhaps, then, it is understandable why the Christian writer Tertullian in around 200 CE could say that the magi were "almost kings."[159]

So now imagine that you have people coming from the group that declares kings in Persia all the way to a land that is under the thumb of Rome, the most powerful rival to the Persians, and declare a toddler is the true ruler of Israel. Those Magi are in fact not only overriding the authority of King Herod, an infamously paranoid tyrant, they are overriding the authority of Caesar Augustus. This is not an idle incident but a declaration of war by the most powerful nation Rome knew. It would be similar if, during the Cold War era, the Soviet Union sent delegates to set up a new governor of Puerto Rico; this would be Cuba's communist takeover squared in the international community. (Perhaps it is no wonder why Herod as well as the people of Jerusalem, according to Matt 2:3, were all filled with dread by the announcements of the Magi.)

In fact, an incident like this happened during the reign of Nero concerning the territory of Armenia, though not with the boldness of the Magi as portrayed in the Gospel. Like Judea, Armenia was a sort of buffer region

---

[156] Herodotus, *Histories* 3.61-74.
[157] M. A. Dandamaev, *Persien unter den erster Achämendian (6. Jh. v. Chr.)*, p. 238.
[158] Strabo, *Geography* 12.2.3, 3.32; Caesar, *De Bello Alexandrino* 66. See also *Cambridge History of Iran*, vol. 3, part 1, pp. 85-86 (on Zoroastrianism patronage from Vologeses I); part 2, pp. 689-691 (on the supreme council), 867-868 (on recognition of Zoroastrianism in 1st century BCE), 904-905 (on taxes for fire temples).
[159] Tertullian, *Contra Marcion* 3.13.

between Rome and its rival, so who ruled this land was a matter of border security. When the Persians appointed a new ruler to this region undesirable to the Romans, Emperor Nero sent in the troops; this lead to a long war and was noted by several historians.[160] Decades earlier, there was a significant diplomatic showdown over the same region and as to who was to appoint its ruler; this did not lead to a war, in part thanks to the negotiating skills of the Roman governor of Syria, but again it made it into several histories from antiquity, including the Jewish historian Josephus.[161]

So if something like this had happened in c. 6 BCE concerning Israel, the response by Rome should have been sending, at the very least, strong condemnations, if not soldiers, and all of Jerusalem should have been afraid (Matt 2:3). Instead, silence, as if nothing happened. And it is not a silence due to a lack of sources for this time as we pretended in the case of eastern records. It is impossible to imagine that Josephus, who had eyewitness records to the life of Herod the Great, would have not mentioned the major diplomatic nightmare the coming of the Magi would have brought. Moreover, the historical reports from the past that are lost to us, such as the writings of another Jewish historian in the first century, Justus of Tiberius, would almost certainly have been preserved by Christians, at least in part, if they contained evidence of the events mentioned in the infancy stories of Jesus. Similarly, some of the writing of Nicholaus of Damascus, a friend of Herod and Josephus's major source for the king's life and final years,

---

[160] Tacitus, *Annals* 13; Dio, *Roman History* 62; Suetonius, *Nero* 57.
[161] Josephus, *Antiquities of the Jews* 18.96-105; Tacitus, *Annals* 2.58; Suetonius, *Caligula* 41.3; *Vitellius* 2.4; Dio, *Roman History* 59.27.3-4.

should have been preserved by the Christians if he talked about this event with the Persian Magi. These unpreserved sources, or even quotes from them, further add to the absence of evidence for this amazing event from taking place, even though it would have been of international importance. One must also wonder how the Gospel of Luke, which styles itself as a historical work, provides no mention of this incredible story. The fact that no source independent of Matthew mentions this story is inexplicable if it really happened, but this silence is exactly what is expected if it never happened. This is a powerful argument from silence because we *expect* to hear something,[162] yet it is not there.

Let's now combine this with everything else that has been established. Western records show no sign of a diplomatic showdown or war expected due to such a trip by the Magi, eastern records show no love for Jesus and instead persecution of his followers, the magi were not interested in astrology or a Jewish king or savior, and our only record for the tale comes from a figure of unknown provenance with unknown sources using terrible historical methodology and relating a story that is filled with the physically impossible. And this is all just considering several verses in Matthew; I haven't even included the story of the Virgin Birth, the contradictions with Luke's story and chronology, and the generally amazing tales that Matthew relates, especially at the death and resurrection of Jesus. We don't even need to consider these objections to see that the

---

[162] Cf. J. P. Moreland and William Lane Craig, *Foundations for a Christian Worldview*, pp. 156-157; Richard Carrier, *Proving History: Bayes's Theorem and the Quest for the Historical Jesus*, pp. 117-119.

historical record is decisively against Matthew's account of the Star being factually correct.

On the other hand, we can see what would have more likely inspired the story, and it has far less to do with events in 6 BCE. There has been a considerable amount of scholarship directed at this question, though not all possibilities are as likely as others, and they may all be to some extent incomplete. However, with the complex nature of any skillfully written story, there are likely several influences, and we need to try and see what were the prospective items that the original audience for the Gospel would have been able to connect to and appreciated. We also do not expect an author to simply copy another story and make small changes, but rather the author communicates their message by what they borrow *and* by what they change.[163] As such, it is necessary to consider several potential links to uncover what the Evangelist was doing.

One idea taken from the historical record concerns another journey of magi, this time though to Emperor Nero in Rome. This event took place in the same year as Halley's comet (66 CE), an eastern potentate comes to receive kingship from Nero, and the new king and his parade of magi return to their homeland via another route (cf. Matt 2:12). Entertained as a possibility by the famous Catholic biblical scholar Raymond Brown and others before him, the most recent treatment is presented by independent scholar Rod M. Jenkins.[164] However, this hypothesis seems to be too weak to explain

---

[163] Thomas L. Brodie, *The Birthing of the New Testament: The Intertextual Development of the New Testament Writings*, pp. 43-49.
[164] Raymond Brown, *Birth of the Messiah*, p. 611; Rod M. Jenkins, "The Star of Bethlehem as the Comet of AD 66", *Journal of the British Astronomical Association* 114, 6 (June 1999): 336-343.

the story. For instance, it does not clarify why the Star was seen "at the rising", let alone any of the miraculous movements of the Star. When it comes to the historical record, none of the ancient sources connect the comet to the arrival of the eastern king (Tiridates) to the court of Nero, the comet had disappeared months before Tiridates's arrival in Rome, and the comet had nothing to do with the ascension of the new king or any prophecies. The only interesting connection to Matthew's story, that Tiridates and his magi left for home by another route, is trivial; besides, there is already an internal narrative logic to why the Magi left by another way: to avoid Herod, a reason having nothing in common with Tiridates's return home. It may have been difficult then for the readers of Matthew to see this minor connection as it would require memory of the coming and goings of kings long dead, not to mention that a comet happened to be seen in that year. With the additional symbolic problems of emulating a story with infamous Nero as the bestower of power along with the evil sign of the comet, there are other possibilities to consider for how the story of the Star of Bethlehem was crafted by the Evangelist or his sources.

Returning focus to the Star itself, one hypothesis put forward first by the early critical New Testament scholar David Strauss in 1835 has us pay attention to one of the most important national myths written for Rome, the *Aeneid* by the poet Virgil during the reign of Caesar Augustus.[165] Early on in the tale, the last surviving Trojans in their war with the Greeks pray for a sign to shepherd them to a new homeland. Suddenly then a blazing shooting star, gliding over the highest housetops,

---

[165] David Straus, *Das Leben Jesu kritisch bearbeitet*, p. 248.

becomes their guide.¹⁶⁶ The star in question has properties like a comet, which is probably because Virgil is incorporating the 'star of Caesar' into his legendary flattery of the Romans and the new emperor; the comet seen at the games of Julius Caesar after his death was interpreted as the soul of the dictator rising up to heaven. More importantly, however, the star seems to be Venus as the morning star and following in an older tradition of Venus as a star guiding her son (Aeneas) to Italy.¹⁶⁷ Considering that the Star of Bethlehem was also supposed to have been a morning star (see the translation in this book for details), then this makes for some dense links between the famous star of Aeneas in a story well-known in the Roman Empire and the story of the Christmas Star in Matthew: a morning star, one that miraculously guides its followers while in the atmosphere, and it leads to a new king(dom) in the West. The myth of Aeneas being led by the morning star was old in the days of the Evangelist, and Virgil's narrative was widely disseminated throughout the empire, so the parallels here would have been recognizable to the intended audience. There is also the notable reversal between the stories: while in Virgil the star guides Aeneas and leads to him coming to power in Italy, for Matthew the Star leads the Magi for the purposes of worshipping another. The reversal is then filled with meaning: the Messiah is the power to follow, even over the authority of other kings and Rome, that the infant Jesus is more important than the founding warrior Aeneas. Worship of the Son of God is paramount compared to earthly sovereigns. While Virgil's story, or

---

[166] Virgil, *Aeneid* 2.687-711.
[167] Servius, *In Virgilii Aeneidos* 1.382; Michael Paschalis, *Virgil's Aeneid: Semantic Relations and Proper Names*, pp. 94-95.

other tales of guiding stars in classical sources, cannot explain the entirety of the creation of the Star of Bethlehem, it is one of several good possibilities for what inspired the tale better known to modern Christians.

Of those other possibilities, there is a particular hypothesis widely considered by scholars that relies on the source we know Matthew was dependent upon: the Hebrew Bible and its numerous prophetic passages. One of the well-known prophecies in antiquity from the Old Testament is Numbers 24:17, which speaks of a star corresponding to the rise of a king for Israel. "A star shall rise out of Jacob, a scepter out of Israel." This was popularly used by various Jews and sects concerning the messiah[168] and is alluded to in an earlier Christian text, Hebrews 7:14. We already see how Matthew cites Old Testament verses to support his story and how his account "fulfilled prophecy", so another example of Hebrew Scripture providing the foundation of one detail of the story is hardly unlikely, and certainly more likely than believing the tale actually happened. Along with this verse, Isaiah 60 talks about eastern kings coming to the rising light of Israel, bringing gold and frankincense, two of the gifts of the Magi. The word that means 'rising' in Matthew 2, *anatole*, is from the same root of the word referring to the rising star in Num 24:17 and the rising light of Israel mentioned in Isaiah 60 in the Greek

---

[168] Cf. David Instone-Brewer, "Balaam-Laban as the Key to the Old Testament Quotations of Matthew 2", in Daniel M. Gurtner and John Nolland, *Built Upon the Rock: Studies in the Gospel of Matthew*, pp. 207-227; Martin Dibelius, *Botschaft und Geschichte: Gesammelte Aufsätze*, I, p. 42 n. 68; Mussies, "Some Astrological Presuppositions of Matthew 2", pp. 25-27; David Sim, "The Magi: Gentiles or Jews?" *Hervormde Teologiese Studies* 55, 4 (1999): 990-996. See also Josephus, *Jewish Wars* 3.399-404; 6.310-315 who applied the prophecy to the Roman general and later emperor Vespasian.

translation, *anatello*. Such a linguistic link helps persuade us that this is what probably brought the tale together, along with other details. In other words, the story is very explicable as a literary creation, including particular points (what gifts are brought) and the choice of words.

On top of these particular verses, we can see Matthew applying the mold of Moses to that of the Jesus story, especially the embellished stories of Moses's birth that were in existence in the first century. Josephus provides such a version in which a scribe who did fortunetelling discovered that a child of the Israelites would be born and would ruin Egypt. Pharaoh became so afraid of this that he ordered the execution of all male Israelite children. In this same legend, there is also Moses's father-to-be worrying about what to do because of the orders from Pharaoh, but in his sleep God told him that all would be okay and his son would be the deliverer of the Hebrew nation.[169] The parallelism with the Jesus birth story in Matthew is unmistakable, a series of tropes that anyone in the ancient world would have seen. But these are not the only similarities between stories of Moses's birth and Jesus's. As noted earlier, there is also a Samaritan version of Moses's birth of uncertain date that included a star that was interpreted and related to the birth of the savior figure.[170] The form of these stories is in obvious conformity with the form of the story of Jesus and the Star, along with Herod acting the part of mad Pharaoh. While Matthew may not have been dependent upon any of these particular versions of Moses's birth, we see that these were the sorts of fictions invented and prevalent at this time, and

---

[169] Josephus, *Antiquities of the Jews* 2.9, §§ 2-3.
[170] Bowman, *Samaritan Documents*, pp. 287-288.

Matthew appeared to be weaving these themes into his narrative.

Outside the Jewish literature, the pagan writers of this era had stories with many of the same tropes: prophecies of the birth of a new leader or savior, miraculous signs at his birth, attempts on his life by an authority figure with the escape of the infant. These things were told of the heroes Perseus and Jason, the superman Hercules, the prophet Zoroaster, the emperor Augustus, the military leader Alexander the Great, the brothers Romulus and Remus, and many others. When these elements were included in Matthew's story of the Nativity, they were already cliché in the first century and fit into a larger category of an ancient Western mythical hero archetype.[171]

But beyond the use of Hebrew Scripture and similar tales to explain the details of the Nativity story, we can see how the account is truly molded in the way expected of fiction. For example, we hear about the mental states of figures such as Herod, the people of Jerusalem, and the Magi. We have conversations between Herod and the Magi (which were held secretly) as well as between Herod and the scribes. These are all things that could not have been witnessed even by the only potential sources for Matthew—Mary and Joseph. However, this exactly fits what is expected from the omniscient narrator—he knows because he is writing the story. There is also the apparent quote from Scripture about how the Messiah was to be born in Bethlehem, but the prophetic verse from the Gospel is a blending of two different verses, 2 Samuel 5:2 and Micah 5:2. This makes for a rather

---

[171] Cf. Robert A. Segal, *In Quest of the Hero*; Richard Horsley, *The Liberation of Christmas: The Infancy Narratives in Social Context*, pp. 162f.

contrived quotation to get the connection the author wants. Conversely, there is no reason that the priests and scribes would have made up the quotation themselves. Other scholars have noted this oddity and the style of the "prophecy" being much like that of the Evangelist, a strong sign that this was not the product of the chief scribes but Matthew.[172] The narrative also shows a fruitful relationship between the chief priests and Herod, while the historical record indicates these two groups were antagonistic; the scribes would not be at the beck and call of the king.[173] Instead, it strikes one as authorial intent guiding the story rather than historical plausibility.

The story as is also contains two inanities. There is no effort on the part of Herod to have spies or any other escort with the Magi which would have guaranteed the destruction of Jesus in the crib (or manger). It's implausible that Herod was paranoid enough to find a baby a threat, and yet he unreservedly trusted the very people that came to contest his claim to the throne. This becomes a plot hole, and one that is necessary for the story to continue. There is also the plot hole as to why the guiding Star failed to guide the Magi to Bethlehem in the first place and avoid Herod from even knowing about the birth of the Messiah. The trip to Jerusalem by the Magi and making their announcement fulfills the purpose of bringing in a villain into the story and produce drama. Obviously this version of the story with Herod is infinitely more memorable compared to the Star

---

[172] William D. Davies & Dale C. Allison Jr., *A Critical and Exegetical Commentary on the Gospel According to Saint Matthew: Introduction and Commentary*, vol. 1, p. 242.
[173] Brown, *Birth*, p. 188.

alone guiding the Magi to Jesus, but it belies how the story is history rather than a concoction.

In fact, we could treat the story of the Star, Jesus and the Magi's escape, and the plot against Jesus's life, as a *pericope*[174] or singular literary unit and analyze it with the method of determining genre, chronotope—how space and time are used in a story to deliver its meaning to the reader. It is how you know that George R.R. Martin's *A Song of Fire and Ice* is in the same genre as J.R.R. Tolkien's *The Lord of the Rings* rather than a modern history of the Crusades or even a novel by Paul Auster or Jane Austin. While chronotopic analysis is usually done over an entire work, chapter 2 of Matthew can also be an examined by this method; after all, the Nativity story should contain the same message as the rest of the Gospel, and it is fruitful to use chronotopic evaluation to see what sort of literature we are dealing with here.

Using the analysis by Bible and literary scholar Michael Vines concerning the Jewish novels from antiquity,[175] we can apply the same observations and see that they also fit Matthew's famous infancy narrative. In the Jewish novels (i.e. Greek Daniel, Greek Esther, Judith, Tobit, *Joseph and Aseneth*), Vines describes a realistic-apocalyptic chronotope (heavenly salvation but in an earthly, historical setting), which has the features of divine versus human sovereignty, God's response to conflict against his preferred people, and the use of human agents to execute His plans to re-establish His sovereignty, with all events happening not because of the human's own resources or skills but through the power

---

[174] Cf. Albrecht Dieterich, "Die Weisen aus dem Morgenlande", *Zeitschrift für die Neutestamentliche Wissenschaft* 3, 1 (1902): 1-14.
[175] Michael Vines, *The Problem of Markan Genre*.

of God. The expanded stories of Moses mentioned above also conform to this paradigm. Matthew 2 fits the chronotope perfectly: the setting is late-first century Judea under Herod; this is a well-known period of conflict with the land ruled by non-Jewish forces; there is conflict between Jesus as the divine King of the Jews and Herod as the human king of the Jews; God allows Jesus to be recognized as the true sovereign, and he avoids an early death and so defeats the king who dies; the Magi succeed not because of their own skills but through the agency of God using both a guiding Star as well as angelic announcements. The conflict or crisis in Jewish novels is the force in the story, and how the characters resolve it is the crux, namely if they remain loyal to God. The Magi fit this: they are instructed by Herod to return to him, yet instead they listen to a messenger from God, which protects Jesus, and hence the integrity of God's plan and sovereignty.

These are not the only literary features in common between the Jewish novels and Matt 2. The element of the weak defeating the strong is also on stage in both: Jesus, a lowly infant, is able to usurp Herod's claim to royalty, while the Magi, without any use of force, are able to defeat the powerful king's desires. We also have the particular connection to the story of Daniel where a foreign king venerates God, and with the Magi involved. Additionally, we see the defeat of Herod's army as they failed to capture the infant Christ all because of the pious actions of Joseph and the Magi; this is comparable to Judith, who all on her own defeats the army of Nebuchadnezzar using her beauty and piety.

In addition, the Jewish novels use irony that shows the superiority of the faith. In Matt 2, there is the irony of foreigners correctly worshipping God and his Son, while the king of the Jews and the chief priests and scribes fail to do so and instead (later in the Gospel) become the chief rivals to God's Son. We have the subversion of the Jews of Palestine and their leaders, a point that is only more pronounced throughout the rest of the Gospel of Matthew. The conflict with rabbinic Judaism and the crowd that has Jesus's blood upon them when they called for his death (Matt 27:22-5) all shows the reversal inherent in the Gospel's message of the Gentile mission (Matt 28:19). Matt 2 already encapsulates the irony of the entire Gospel, of foreign piety over Jewish hypocrisy. As such, we ought to place this story (and perhaps the entire Gospel of Matthew) into the category of novel. Not history, not critical biography, but edifying, theological fiction as pseudo-historical as were the novels mentioned above. That the elements of the tale can also clearly be derived from the Old Testament overwhelmingly demonstrates the purpose and method of the story's creation, and it has nothing to do with historicity.

So, no matter what prior probability you may have reasoned initially for the truth of the tale (unless you irrationally claim 100% certainty or near it), even if you accept the miraculous and don't think it can affect the probability of the tale (itself unreasonable), the evidence is overwhelming against the historicity hypothesis. However, it is possible to make sense out of how such a story could have come about when examining the only source we know the authors of the Gospels used—the Hebrew Bible—and the literary universe the Evangelists existed in. All this decisively shows that the tale of the Star of Bethlehem is untenable as history, even if you believe in miracles or still claim it was a natural phenomenon. The evidence is extraordinary, but extraordinarily against historical truth in Matt 2:1-12.

At this point I have not considered things such as the Slaughter of the Innocents or other historically questionable aspects of the Nativity in the Gospel of Matthew, leaving that to Jonathan Pearce in his book on the subject amongst other critical books and articles about the Nativity. However, because in this case we can be effectively certain that Matthew has concocted material about Jesus's birth, it follows that this will also lower the chances that any of his other stories about Jesus are likely to be true, especially the infancy narrative. Consequently, this excursion into just one part of the story does in fact affect the plausibility of the rest—if this was made up, then it is more likely other things were also invented.

In conclusion: No, Virginia, there was neither a Star of Bethlehem nor a journey of the Magi. It's a story of significant importance theologically, perhaps, but it has nothing more to do with history than the non-canonical tales of Jesus's infancy and childhood where he makes living clay birds and fights dragons.

APPENDIX: THE TEXT OF MATTHEW

The primary language of the world the New Testament found itself in was Greek, largely because of the expansionist conquests of Alexander the Great centuries before the Gospels were composed. The Romans, though certainly not loathsome of their own tongue, also learned this language, and it became the *lingua franca* of the Empire. Most people who needed to trade would know at least some Greek, and students were taught it in written form, as well as learning their native language, such as Latin or Hebrew. So it is no surprise that the Gospel of Matthew is written in the Greek language, and as such it could reach the widest possible audience. We are also fortunate that the chapters that open the Gospel are reasonably well-preserved in the manuscripts we have, though they are all dated to centuries after when the autograph was composed (very broadly around 100 CE). Among the manuscripts we have, there is little variance besides missing lines or misspellings, the normal and expected situation when all books had to be copied by hand.

Our primary manuscripts are the codices from the fourth century, such as Codex Vaticanus and Codex Sinaiticus. These are known as a part of the Alexandrian text-type, which are some of the oldest texts and they are considered the most reliable. The other major text-type categories are the Western and Byzantine. The latter makes up the greatest number of copies of the New Testament, but they are all younger and less revealing of what the autographs of the Gospels would have looked like. The Western text-type is older than the Byzantine type and covers a wide geographical spread of manuscripts.

However, these manuscripts tend to paraphrase and have textual corruptions. Perhaps the most noticeable difference between the Western and Alexandrian text-types is that the Book of Acts is about 10% longer in the Western version compared to the Alexandrian, and it is uncertain which is closer to the original.

For the most part, textual critics are reliant upon Alexandrian text-types because of their antiquity, their greater uniformity (better controls in copying, so more likely to reflect the earliest versions of the Gospel they had for copying), and less tendency to paraphrase, expand, or otherwise "improve" the texts. The codices mentioned above are written without diacritical marks (dashes and curves that indicate how words are pronounced; e.g. ά, ὰ) and in all capital letters without spaces between words or punctuation marks that we take for granted. Modern critical editions of the Greek New Testament are usually more forgiving: they use upper and lower case as common English users do, they use punctuation marks, and they include diacritical marks.

The Greek text here is that agreed upon by the United Bible Societies (UBS) in their 4th edition of the Greek New Testament. For chapter 2 for the Gospel of Matthew, the USB version of the text does not differ from the older, classic version compiled by the textual critics Brooke Foss Westcott and Fenton John Anthony Hort published in 1881. There is also almost no difference between this edition of the Gospel chapter and the Greek text that was used for the King James Bible, the Textus Receptus, the only difference of note being in verse 9, using ἔστη instead of ἐστάθη—the active and passive forms of the verb ἵστημι (to stop, stand). In the context of the verse, it makes no substantial difference. Fortunately for the interpretation of this story, we have less reason to worry about textual uncertainties and interpolations that in other parts of the Bible, and we can avoid dogmatic debates about what editions of the Greek New Testament should be given theological weight.

As for the English translation, this is my own, though it is informed by the numerous, modern English versions on the market, especially the Revised Standard Version (RSV). However, a few points in the Greek tend to not be translated with respect to the Star of Bethlehem as they make the phrasing clumsy and not terribly informative. However, I will try to include these points since they have had an effect on my analysis of the Star. There is also some uncertainty about the phrase ἐν τῇ ἀνατολῇ, which needs its own discussion, but my understanding here is sufficiently in line with other translations and research to not be unsoundly rendered as "at the rising of the sun".[176] This can be compared to how scholars translate the phrase in the works of the historian and essayist Plutarch (*Moralia* 284e9) and the astronomer Ptolemy,[177] both of whom were broadly contemporary with the Evangelist. Other common translations include "in the East" and "at its rising", with the latter more often found in modern translations.

---

[176] Cf. Roberts, *The Star of the Magi*, pp. 120-1. Roberts contacted two translators of ancient Greek astrological texts for their opinions on the phrase in question.

[177] Johan Ludvig Heiberg, *Claudii Ptolemaei opera quae exstant omnia I: Syntaxis mathematica, Part 2*, p. 595; G. J. Toomer, *Ptolemy's Almagest*, p. 638.

1 Τοῦ δὲ Ἰησοῦ γεννηθέντος ἐν Βηθλέεμ τῆς Ἰουδαίας ἐν ἡμέραις Ἡρῴδου τοῦ βασιλέως, ἰδοὺ μάγοι ἀπὸ ἀνατολῶν παρεγένοντο εἰς Ἱεροσόλυμα

2 λέγοντες, Ποῦ ἐστιν ὁ τεχθεὶς βασιλεὺς τῶν Ἰουδαίων; εἴδομεν γὰρ αὐτοῦ τὸν ἀστέρα ἐν τῇ ἀνατολῇ καὶ ἤλθομεν προσκυνῆσαι αὐτῷ.

3 ἀκούσας δὲ ὁ βασιλεὺς Ἡρῴδης ἐταράχθη καὶ πᾶσα Ἱεροσόλυμα μετ' αὐτοῦ,

4 καὶ συναγαγὼν πάντας τοὺς ἀρχιερεῖς καὶ γραμματεῖς τοῦ λαοῦ ἐπυνθάνετο παρ' αὐτῶν ποῦ ὁ Χριστὸς γεννᾶται.

5 οἱ δὲ εἶπαν αὐτῷ, Ἐν Βηθλέεμ τῆς Ἰουδαίας: οὕτως γὰρ γέγραπται διὰ τοῦ προφήτου:

6 Καὶ σύ, Βηθλέεμ γῆ Ἰούδα, οὐδαμῶς ἐλαχίστη εἶ ἐν τοῖς ἡγεμόσιν Ἰούδα: ἐκ σοῦ γὰρ ἐξελεύσεται ἡγούμενος, ὅστις ποιμανεῖ τὸν λαόν μου τὸν Ἰσραήλ.

7 Τότε Ἡρῴδης λάθρᾳ καλέσας τοὺς μάγους ἠκρίβωσεν παρ' αὐτῶν τὸν χρόνον τοῦ φαινομένου ἀστέρος,

8 καὶ πέμψας αὐτοὺς εἰς Βηθλέεμ εἶπεν, Πορευθέντες ἐξετάσατε ἀκριβῶς περὶ τοῦ παιδίου ἐπὰν δὲ εὕρητε ἀπαγγείλατέ μοι, ὅπως κἀγὼ ἐλθὼν προσκυνήσω αὐτῷ.

9 οἱ δὲ ἀκούσαντες τοῦ βασιλέως ἐπορεύθησαν, καὶ ἰδοὺ ὁ ἀστὴρ ὃν εἶδον ἐν τῇ ἀνατολῇ προῆγεν αὐτοὺς ἕως ἐλθὼν ἐστάθη ἐπάνω οὗ ἦν τὸ παιδίον.

10 ἰδόντες δὲ τὸν ἀστέρα ἐχάρησαν χαρὰν μεγάλην σφόδρα.

11 καὶ ἐλθόντες εἰς τὴν οἰκίαν εἶδον τὸ παιδίον μετὰ Μαρίας τῆς μητρὸς αὐτοῦ, καὶ πεσόντες προσεκύνησαν αὐτῷ, καὶ ἀνοίξαντες τοὺς θησαυροὺς αὐτῶν προσήνεγκαν αὐτῷ δῶρα, χρυσὸν καὶ λίβανον καὶ σμύρναν.

12 καὶ χρηματισθέντες κατ' ὄναρ μὴ ἀνακάμψαι πρὸς Ἡρῴδην, δι' ἄλλης ὁδοῦ ἀνεχώρησαν εἰς τὴν χώραν αὐτῶν.

1 After Jesus was born in Bethlehem of Judea in the days of Herod the King, behold, Magi from the East came into Jerusalem

2 saying, "Where is the one born King of the Jews, for we saw his Star at the rising of the sun, and we have come to pay homage to him.

3 Hearing this, King Herod was disturbed and all of Jerusalem with him,

4 and bringing together all the chief priests and scribes of the nation he began enquiring of them where the Messiah was to be born.

5 They told him, "In Bethlehem of Judea, as it is written in the writings of the prophets:

6 'And you, Bethlehem of the land of Judah, who by no means is least of the rulers in Judah, from you shall come a leader who will nourish my people, the Israelites.'"

7 Then Herod secretly invited the Magi and discovered from them the time of the appearance of the Star.

8 And he sent them to Bethlehem and said, "Depart and search diligently for the child, and when you find him bring me word so that I may also come revere him."

9 After hearing the king, they left for their journey, and lo! The Star which they had seen at the rising of the sun led them on until it arrived and stood over the place where the child was.

10 Upon seeing the Star, they greatly rejoiced with great joy.

11 And entering into the house they saw the child with Mary his mother, and prostrated themselves and paid him homage, and opening their treasures they gave him gifts: gold, frankincense, and myrrh.

12 And being commanded in a dream not to return to Herod, they traveled back to their country through another route.

GLOSSARY

Achaemenid Empire: the first Persian Empire. Established by Cyrus the Great in his defeat of the Babylonian Empire in the 6th century BCE, it became the largest empire in the world, stretching from Afghanistan to Egypt and into Europe. Cyrus is called the messiah in the Bible (Isaiah 45:1) and helped in rebuilding the Jewish temple (2 Chronicles 36; Ezra 1). The empire fell because of the military conquests of Alexander the Great in the 330s BCE.

Ascendant: one of the **cardinal points** of a **horoscope**. This is the location of the eastern horizon at a certain degree of an astrological sign.

Aspect (astrology): the special angular relations between **signs** and/or **planets** in a **horoscope**. The most common aspects used in antiquity are **conjunction** (0° separation), sextile (60° separation), square (90° separation), **trine** (120° separation), and opposition (180° separation). The angular separations need not be exact to count for an aspect, but the size of the margin around these angles (called orbs) is debated by astrologers.

Astrology: the divinatory technique of predicting things based on the arrangements of the stars and **planets**. One of the chief tools used for such predictions is the **horoscope**. Most practicing astrologers today are not astronomers and use tables or computers to compose a horoscope. Starting in ancient Babylonia, astrology has gone from a scientific endeavor by many competent researchers to a pseudoscience existing outside of standard astronomical research. Johannes Kepler (1571-1630) is often considered the last great scientific astrologer. In well-formed tests of astrology, professionals show an accuracy level no better than chance and a level of agreement almost no better than chance.

Astronomy: the study of the universe beyond the Earth. This includes other **planets**, stars, galaxies, and the entire universe itself. The subject includes the use of physics, chemistry, geology, and biology. With prehistoric roots, the science of astronomy began in Babylonia, advanced under the Greeks, Romans, and medieval Arabs, and had a strong effect in igniting the Scientific Revolution. Popular areas of research include planetary formation, stellar evolution, and cosmology (origins, structure, and fate of the universe). The primary tools of the modern astronomer are telescopes that use various wavelengths of light (visible, infrared, radio, etc.) and computers, both for data analysis and for modeling. Planetary scientists also use probes that travel to solar system bodies, including the Cassini spacecraft around Saturn and the rover Curiosity on Mars.

Cardinal points: the four compass points on a **horoscope**. Those points are the **ascendant** (Asc), the descendant (Des, 180° opposite the ascendant), the **midheaven** (MC), and the *imum coeli* (IC, 180° opposite the midheaven). These points are important for the interpretation of a horoscope.

Chronotope: a tool in literary theory to describe the genre of a work. Devised by Mikhail Bakhtin in the 1930s, it looks at how time (Gk. *chronos*) and space (Gk. *topos*) are represented in a narrative. Chronotope can be established by the setting, the plot, the actions of characters, how events change characters, the relation of the narrator's voice to the story, etc. As Bakhtin describes: "Time, as it were, thickens, takes on flesh, becomes artistically visible; likewise, space becomes charged and responsive to the movements of time, plot and history." The analysis of chronotope is done by comparison with the literary context the work finds itself in and how they establish their respective chronotopes; in the case of the Gospels, one considers the literature of that time that was influential (i.e. the Old Testament, the works of Homer, popular novels, etc.) and how they used chronotope (i.e. Homer's use of sea voyages and divine intervention, the ancient Greek novels' use of the element of chance, etc.).

Comet: from the Latin for hair (*coma*), these are balls of water ice, ammonia, carbon dioxide, as well as rock and dust and thus called "dirty snowballs" in popular literature. The nucleus of a comet will begin to boil off gas and dust when heated up on its approach to the Sun, producing the tail (or tails) of the comet, the 'hair' of the star. Often two tails can be seen, one composed of dust and the other of gas; the latter tail is most strongly pushed by the solar winds and points away from the Sun. Tails can reach lengths over 100 million kilometers long. The dust of comets remains behind and can produce meteor showers for Earth-based viewers at certain times of the year. Comets are theorized to come from a sphere of icy bodies surrounding the Sun in what is called the Oort cloud; gravitational perturbations from other objects in space occasionally send a comet falling towards the inner solar system. Because they are orbiting the Sun, most comets are periodic in their appearances, but gravitational effects by the planets of the solar system, Jupiter especially, can change the orbit and the period. Since antiquity, comets have almost universally been interpreted as frightening omens.

Conjunction: an astrological **aspect** when two objects have the same **right ascension**. In a **horoscope**, they would appear in the same spot. Conjunctions can happen between stars and **planets**, between planets, etc.

Constellation: a collection of stars that imaginatively look like or represent something. Different cultures have different constellations, but the official Internal Astronomical Union (IAU) system has 88 constellations, most of which come from Greco-Roman traditions.

Cuneiform: one of the oldest forms of writing, emerging in the 4th millennium in Sumer. The script is formed using wedge shapes and was used by the Babylonians, Assyrians, and other Ancient Near Eastern cultures. It fell out of use progressively in the first millennium BCE and disappeared by the 2nd century CE. It was unlike the modern alphabet where each letter had a particular phonetic (sound) value but instead some were words, some syllables, and some consonants. After it went out of use, cuneiform was not deciphered until the 19th century. One of the oldest extant works of fiction, the *Epic of Gilgamesh*, was written in cuneiform.

Declination: a part of the (celestial) equatorial coordinate system equivalent to latitude on Earth. The zero line of declination is called the celestial equator and is the projection of Earth's equator into space. Above and below the celestial equator gives positive and negative declinations respectively, ranging between -90° and +90°.

Epicycle: literally a cycle on top of another cycle. A **planet** on an epicycle follows a circular orbit around a point which is itself moving in a circle (called a deferent) around another point. A planet doing this could look to an observer who is on or near the center of the deferent as if it slowed, reversed direction, slowed again, and returned moving the way it did before. The system of epicycles was the method primarily used to produce a geocentric model of the solar system.

Form criticism: the method of labeling literary units and determining their functions and origins. For much of 20th century biblical studies, this meant determining the oral traditions upon which the Bible authors relied on to produce their stories. Some of the most famous form critics of the New Testament include Rudolf Bultmann, Martin Debelius, and Karl Ludwig Schmidt.

Great conjunctions: the **conjunctions** of Jupiter and Saturn, which take place approximately every 20 years. Successive great conjunctions take place about 120° apart on a horoscope and thus remain in the same **trine**. When such a conjunction takes place in a different trine than the previous conjunction, it is called a greater conjunction; these take place approximately every 240 years. When a great conjunction first takes place in the fiery trine, it is called the greatest conjunction. Without considering the effects of **precession**, which moves the background stars as seen by an Earth-based observer and thus the location of the constellations, greatest conjunctions take place about 960 years apart or nearly a millennium. With the effects of precession, the period is about 800 years and will have 40 great conjunctions. Astrologers believed these conjunctions had effects on world affairs. This astrological theory developed during the **Sassanid period** in the Middle East (3rd-7th centuries CE).

Horoscope (Western): a map of the **zodiac signs**, the **planets**, and the **cardinal points**. The relative positions of these various bodies are what lead to the interpretation of the horoscope. The earliest horoscopes known come from the 5th century BCE in Babylonia, but the Western form described above did not come to be until the 2nd century BCE in Hellenistic Egypt. The oldest extant treatise on horoscopic astrology comes from the early 1st century CE.

Magus/Magi: priest(s) in the **Zoroastrian** religion. The name seems to derive from the name of the religious caste of the Medes in what is now western Iran. Through the Greek *magoi* and Latin *magi*, the title would become the modern word 'magic.'

Meteor: the visible streak of light in the sky caused by a falling rock from space. Meteors can be seen at any time of the night in any direction, but during meteor showers they will appear to come from a certain part of the sky. Meteor flashes can last from a second to over a minute, though usually they are sudden appearances. If any part of the meteor reaches the ground, it becomes known as a meteorite.

Midheaven (*medium coeli*, MC): one of the **cardinal points** of a **horoscope**. It is the location of the highest part of the **zodiac** seen at the time and location of the horoscope's object of consideration. The angle between the midheaven and **ascendant** (the horizon) is usually less than 90°, but often in ancient horoscopes the angle is simply treated as a right angle.

Nova: a stellar explosion that happens when gas from another star is accreted on to the surface of a white dwarf star. Gravitational pressure on the gas causes fusion, and there is a fast explosion on the surface of the white dwarf. To an Earth-based observer, it appears as if a star has just appeared (hence the Latin word for 'new', *nova*, used for the object). In a short while, usually days, the nova will dim and become invisible to the naked eye. Some novae are recurrent and explode periodically.

Parthian Empire: also called the Arsacid Empire, it developed out of uprisings against the dynastic empires created by Alexander's generals, especially that of Seleucus. Established by Arsaces I of Parthia in 247 BCE, it would become the chief rival power to Rome in the East. It finally fell in the early 3rd century CE and replaced by another Persian empire, that of the **Sassanids**.

Pericope: a set of clauses or verses that can form a literary unit or thought. For much of the 20th century, Bible scholars employing **form criticism** used pericopes (or pericopae) to determine different traditions and sources for the Gospels.

Planet (classical world): from the Greek for wanderer, one of seven objects that are star-like in appearance but move against the background **constellations**. In antiquity, the seven objects were (in the common order given as their relative distance from the Earth): Moon, Mercury, Venus, Sun, Mars, Jupiter, and Saturn.

Precession of the equinoxes: the slow movement of the location of the Sun in the **zodiac** on the first day of spring (the vernal equinox). The effect is due to the "wobble" of the Earth, similar to how a top wobbles or precesses, which is caused by the gravitational forces between the Earth and the Moon and Sun. Precession is cyclical with the north pole of the Earth drawing out a circle in the sky over a period of about 26,000 years. The discovery of the effect is credited to Hipparchus of Rhodes in the 2nd century BCE, and the oldest extant discussion of precession comes from Claudius Ptolemy in the 2nd century CE.

Right ascension: a part of the (celestial) equatorial coordinate system equivalent to longitude on Earth. It measures from the vernal equinox towards the east. Instead of using degrees, it is measured by using 24 hours for a complete circle instead of 360°.

Sassanid Empire: the last Persian empire before the rise of Islam. Founded by Ardashir I in the 3rd century CE, the **Zoroastrian** religion received significant state support while occasionally suppressing other faiths. Christianity would be recognized in 409/410 by King Yazdegerd I, though the relation between the empire and the Christians would swing back and forth between acceptance and suppression. The Sassanid Empire would continue to rival Rome and the Byzantine Empire for eastern territories until its fall during the Arab conquests of the 7th century.

Sign (astrology): similar to a **constellation**, one of 12 regions of the **zodiac** 30° across. The names of the signs come from the zodiac constellations. Signs do not necessarily line up with the visible constellations because of the precession of the equinoxes and because constellation are not of equal size. In what is called a tropical zodiac, the defining point of the sign Aries is the location of the vernal equinox (where the Sun is on the first day of spring), and from there all the other signs get their locations. Signs have various attributes, such as gender and element.

Supernova: the most colossal form of stellar explosion. The two major types are Type Ia and Type II. The former happens when a white dwarf star accretes enough material from another star to reach the Chandrasekher limit (~1.4 solar masses) and collapse, releasing a huge amount of energy. Type II supernovae happen when a massive star runs out of fuel for fusion and collapses, the infalling gas bounces off the core of the star, and then explodes outward. Supernova explosions can be more luminous than an entire galaxy of stars. Seen from the Earth, it would appear as a particularly bright star, which over time would dim and disappear. The last naked-eye supernova in the Milky Way was seen in 1604 (Kepler's nova).

Trine/Trigon/Triplicity: three **signs** that are 120° apart in a **horoscope** and are assigned one of the four classical elements. The fiery trine, for example, consists of Aries, Leo, and Sagittarius. A trine **aspect** refers to two **planets** that are about 120° apart or in two signs in trine aspect (i.e. Jupiter in Aries, Mars in Leo). A trine aspect is usually considered positive by astrologers.

White dwarf: the remnant core of a star after all the gas that could be fused is exhausted and the compact remnant remains. Consisting largely of carbon, oxygen, and heavier elements, white dwarves have an extreme density of about 100,000 times that of lead. The gravitational force keeps the star so extremely dense, while it is the quantum mechanical effect of electron degeneracy pressure that prevents it from collapsing. If a white dwarf receives more mass, such as gas from another star, it can become so massive it overcomes the degeneracy pressure and collapse into a denser object, such as a neutron star or black hole and in the process creates a **supernova**. White dwarves are very dim, and none are seen by naked-eye observers on Earth. An example of a white dwarf is the companion star to the brightest star in the sky, Sirius; this dwarf is known as Sirius B.

Zodiac: the ring of **constellations** or **signs** in which the classical **planets** move through. Traditionally there are 12 constellations, most of them animals (hence the *zoo* in zodiac). The Sun will move through the entire zodiac once a year. The zodiac was identified first in the 7th century BCE by the Babylonians.

Zoroastrianism: a religion traditionally founded by the prophet Zarathustra (to the Greeks and Romans 'Zoroaster') in the 2nd millennium BCE, in which there are two chief gods, one good (Ahura Mazda, later Ohrmazd) and one evil (Angra Mainyu, later Ahriman). Participants in the religion take part in beliefs of a cosmic battle between these gods, including a final confrontation and the destruction of Ahriman. In each of the last millennia, a savior figure called the Saoshyant would appear, the last one bringing about the final revelation and victory for Ahura Mazda. It was a major religion in Persia, and its priests were the **magi**. The Avesta is the primary collection of sacred literature, consisting of liturgical works (*Yasna*), hymns to particular gods (*Yasht*), catalogues of evil spirits and the maladies (*Vendidad*), and more. The belief in an evil anti-god, a savior figure, and a final judgment of the living and dead all likely influenced Judaism during the **Achaemenid** Period. There are perhaps as many as one or two million followers of this faith today. In modern times, the best known Zoroastrian is probably the late Farrokh Bulsara, better known as Freddie Mercury.

BIBLIOGRAPHY

Adair, Aaron. "Science, Scholarship, and Bethlehem's Starry Night", *Sky & Telescope* 114, 6 (Dec 2007): 26-29
— "The Star of Christ in the Light of Astronomy", *Zygon: Journal of Science & Religion* 47, 1 (March 2012): 7-29.
Alexander, Gerhard. *Apologie; oder Schutzschrift für die vernünftigen Verehrer Gottes*. Frank am Main: Insel Verlag, 1972.
Asimov, Issac. *The Planet that Wasn't*. Garden City, NY: Doubleday, 1976.
Baldwin, Barry. *Suetonius*. Amsterdam: A. M. Hakkert, 1983.
Banos, George. "What the star of Bethlehem Uranus", *Astronomy Quarterly* 3, 12 (1980): 165-168.
Barnes, Timothy. "The Date of Ignatius", *The Expository Times* 120, 3 (2008): 119-130.
Barton, Tamsyn. *Ancient Astrology*. London: Routledge, 1994.
Barton, John. Muddimen, John. *The Oxford Bible Commentary*. Oxford: Oxford University Press, 2001.
Beck, Roger. "Thus Spake Not Zarathustra: Zoroastrian Pseudopigrapha of the Greco-Roman World" in *A History of Zoroastrianism*, Mary Boyce, ed. (Leiden: Brill, 1975-1991) Vol. 3, pp. 491-565
— *A Brief History of Astrology*. Malden, MA: Blackwell Publishing, 2007.
Boll, Franz. "Der Stern der Weisen", *Zeischrift für die neutestamentlich Wissenschaft*, 18, Jahrg., Heft 1/2 (1917): 40-48.
Bowman, John. *Samaritan Documents: Relating to their History, Religion, and Life*. Pittsburgh: Pickwick Press, 1977.
Boyce, Mary (ed.). *A History of Zoroastrianism*. Leiden: Brill, 1975-1991.
Bradley, K. R. *Suetonius' Life of Nero: An Historical Commentary*. Bruxelles: Latomus, 1978.
Brodie, Thomas L. *The Birthing of the New Testament: The Intertextual Development of the New Testament Writings*. Sheffield: Sheffield University Press, 2004.

Brown, Raymond E. *The Birth of the Messiah: A Commentary on the Infancy Narratives in the Gospels of Matthew and Luke.* New York: Random House, 1993.

Burnett, Andrew. Amandry, Michael. Alegre, P. P. Ripollés. Butcher, Marguerite Spoerri. *Roman Provincial Coinage.* London: British Museum Press, 1998.

The Cambridge History of Iran. Cambridge: Cambridge University Press, 1968-1991.

Carrier, Richard. *Proving History: Bayes's Theorem and the Quest for the Historical Jesus.* Amherst, NY: Prometheus Books, 2012.

Caspar, Max. *Kepler.* New York: Dover, 1993.

Clark, David H. Parkinson, John H. Stephenson, F. Richard. "An Astronomical Re-Appraisal of the Star of Bethlehem—A Nova in 5 BC", *Quarterly Journal of the Royal Astronomical Society* 18 (1977): 443-449.

Clarke, Arthur C. "The Star", *Infinity Science Fiction* 1, 1 (1955): 120–127.

Clow, Barbara Hand. *Chiron: Rainbow Bridge between the Inner and Outer Planets.* St. Paul: Llewellyn, 1987.

Cullen, Christopher. "Can We Find the Star of Bethlehem in Far Eastern Records?" *Quarterly Journal of the Royal Astronomical Society* 20 (1979): 153-159.

Dandamaev, M. A. *Persien unter den erster Achämendian (6. Jh. v. Chr.).* Wiesbaden: Reichert, 1976.

Davies, William D. Allison, Dale C. *A Critical and Exegetical Commentary on the Gospel According to Saint Matthew: Introduction and Commentary.* Edinburgh: T. & T. Clark, 1988-1997.

de Jong, Albert. *Traditions of the Magi: Zoroastrianism in Greek and Latin Literature.* Leiden: Brill, 1997.

dei Rossi, Azahiah ben Moses. Weinberg, Joanna. *The Light of the Eyes.* New Haven: Yale University Press, 2001.

Dibelius, Martin. *Botschaft und Geschichte: Gesammelte Aufsätze.* Tübingen: Mohr, 1953-1956.

Dieterich, Albrecht. "Die Weisen aus dem Morgenlande", *Zeitschrift für die Neutestamentliche Wissenschaft* 3, 1 (1902): 1-14.

Dione, R. L. *God Drives a Flying Saucer*. New York: Bantam Books, 1968.

Dodge, Bayard. *The Fihrist of al-Nadim: A Tenth-Century Survey of Muslim Culture*. New York: Columbia University Press, 1970.

Downing, Barry H. *The Bible and Flying Saucers*. New York: Avon, 1970.

Elwell, Douglas A. *Planet X, the Sign of the Son of Man, and the End of the Age*. Crane, MS: Defender Publishing Group, 2011.

Faracovi, Ornella Pompeo. *Gli oroscopi di Cristo*. Venezia: Marsilio, 1999.

Ferrari, Leo. "Astronomy and Augustine's Break with the Manichees", *Revue des Études Augustiniennes* 19 (1973): 263–276.

Ferrari d'Occhieppo, Konradin. "The Star of Bethlehem", *Quarterly Journal of the Royal Astronomical Society* 19 (1978): 517-520.

— "Neue Argumente zu Aufgang und Stillstand des Sterns in der Magierperikope Matthaüs 2, 1-12", *Sitzungsberichte der Österreichische Akademie der Wissenschaften*, Abteilung II. Mathematische, Physikalische und Technische Wissenschaften, 206 (1997): 317-344.

Freitag, Ruth S. *The Star of Bethlehem: A List of References*. Washington: Library of Congress, 1979.

Frisch, Christian. Joannis *Kepleri Astronomi Opera Omnia*. Francofurti a.M. et Erlanga: Heyder & Zimmer, 1858-1871.

Gardiner, Martin. *Urantia: The Great Cult Mystery*. Amherst, NY: Prometheus Books, 1995.

Giberson, Karl. Collins, Francis. *The Language of Science and Faith: Straight Answers to Genuine Questions*. Downers Grove, IL: Intervarsity Press, 2011.

Gignoux, Philippe. "L'Inscription de Kartir à Sar Mashad", *Journal Asiatique* 256 (1968): 387-418.

Gingerich, Owen. Hoskin, Michael. Hughes, David. Birdsall, J. Neville. "Review Symposium: The Star of Bethlehem", *Journal for the History of Astronomy* 33, 4, (2002): 386-394.

Goldstein, Bernard R. Pingree, David E. *Levi ben Gerson's Prognostication for the Conjucntion of 1345*. Transactions of the American Philosophical Society, vol. 80, pt. 6. Philadelphia: American Philosophical Society, 1990.

Goodacre, Mark. *The Case Against Q: Studies in Markan Priority and the Synoptic Problem*. Harrisburg, PA: Trinity Press International, 2002.

Gundel, Wilhelm. Schott, Siegfried. *Dekane und Dekansternbilder: Ein Beitrag zur Geschichte der Sternbilder der Kulturvölker*. Glückstadt und Hamburg: J.J. Augustin, 1936.

Heiberg, Johan Ludvig. *Claudii Ptolemaei opera quae exstant omnia I: Syntaxis mathematica*. Lipsiae : Teubner, 1898.

Ho Peng Yoke. "Ancient and Medieval Observations of Comets and Novae in Chinese Sources", *Vistas in Astronomy*, vol. 5 (1962): 127-255.

Horsley, Richard. *The Liberation of Christmas: The Infancy Narratives in Social Context*. New York: Crossroad, 1989.

Hughes, David W. *The Star of Bethlehem: An Astronomer's Confirmation*. New York: Walker, 1979.

Humphreys, Colin. "The Star of Bethlehem—a Comet in 5 BC—and the Date of the Birth of the Christ", *Quarterly Journal of the Royal Astronomical Society*, 32 (1991): 389-407.

Hunger, Hermann. *Astrological Reports to Assyrian Kings*. Helsinki: Helsinki University Press, 1992.

Hunger, Hermann. Parpola, Simo. "Bedechungen des Planeten Jupiter durch den Mond", *Archiv für Orientforschung*, 29/30 (1983/4): 46-49.

Hunger, Hermann. Stephenson, F. Richard. Walker, C. B. F. et al. *Halley's Comet in History*. London: British Museum Publications, 1985.

Instone-Brewer, David. "Balaam-Laban as the Key to the Old Testament Quotations of Matthew 2", in *Built Upon the Rock: Studies in the Gospel of Matthew*, Daniel M. Gurtner and John Nolland, eds. (Grand Rapids, MI : William B. Eerdmans Pub., 2008), pp. 207-227.

Jenkins, Rod M. "The Star of Bethlehem as the Comet of AD 66", *Journal of the British Astronomical Association* 114, 6 (June 1999): 336-343.

Keller, Werner. *The Bible as History*. New York: Barnes & Noble, 1995.

Kennedy, Edward S. "Comets in Islamic Astronomy and Astrology", *Journal of Near Eastern Studies* 16, 1 (1957): 44-51.

Kennedy, Edward S. Pingree, David. *The Astrological History of Masha'allah*. Cambridge, MA: Harvard University Press, 1971.

Kepple, George Robert. Sanner, Glen W. *The Night Sky Observer's Guide*. Richmond, VA: Willmann-Bell, 1998.

Kronk, Gary W. *Cometography: A Catalog of Comets*. Cambridge: Cambridge University Press, 1999.

Kidger, Mark R. *The Star of Bethlehem: An Astronomer's View*. Princeton: Princeton University Press, 1999.

Koch, Dieter. *Der Stern von Bethlehem*. Frankfurt: Verl. der Häretischen Blätter, 2006.

Koch-Westenholz, Ulla. "The Astrological Commentary *Šumma Sin ina tamurtišu* Table 1" in *La Sciences des Cieux: Sages, Mages, Astrologues*, Rika Gyselen, ed. (Bures-sur-Yvette: Groupe pour l'étude de la civilisation du Moyen-Orient, 1999), pp. 149-166.

Kühnöl, Christian Gottlieb. *Evangelium Matthaei*. Leipzig: Barth, 1807.

Landau, Brent Christopher. "The Sages and the Star-Child: An Introduction to the *Revelation of the Magi*, an Ancient Christian Apocryphon", Th.D. Thesis: Harvard University, 2008.

MacDonald, George. "The Pseudo-Autonomous Coinage of Antioch", *The Numismatic Chronicle* 4 (1904): 105-135.

Martin, Ernest L. *The Star that Astonished the World*. Portland: ASK Publications, 1991.

McDowell, Al. *Uncommon Knowledge: New Science on Gravity, Light, the Origin of Life, and the Mind of Man*. Bloomington, IN: AuthorHouse, 2009.

McGrew, John. McFall, Richard. "A Scientific Inquiry into the Validity of Astrology", *Journal of Scientific Exploration* 4, 1 (1990): 75-83.

Miller, Robert J. *The Complete Gospels: Annotated Scholars Version*. San Francisco: Harper, 1992.

Mobberley, Martin. *It Came from Outer Space Wearing an RAF Blazer!: A Fan's Biography of Sir Patrick Moore*. New York: Springer, 2013.

Molnar, Michael R. *The Star of Bethlehem: The Legacy of the Magi*. New Brunswick: Rutgers University Press, 1999.

Moore, Patrick. *The Star of Bethlehem*. Bath: Canopus Publ., Ltd., 2001.

Moreland, J. P. Craig, William Lane. *Foundations for a Christian Worldview*. Downers Grove, IL: InterVarsity Press, 2003.

Mussies, Gerard. "Some Astrological Presuppositions of Matthew 2: Oriental, Classical and Rabbinical Parallels", in *Aspects of Religious Contact and Conflict in the Ancient World*, ed. Peter Willem van der Horst (Ultrecht: Faculteit der Godgeleerdheid Universiteit Utrecht, 1995), pp. 25-44.

Nanninga, Rob. "The Astrotest: A Tough Match for Astrologers," *Correlation* 15, 2 (1996): 14-20.

Neugebauer, Otto. *A History of Ancient Mathematical Astronomy*. Berlin: Springer-Verlag, 1975.

Newman, Robert. "The Star of Bethlehem: A Natural-Supernatural Hybrid?" *Interdisciplinary Biblical Research Institute* 2001, pp. 1–16.

Nigosian, Solomon. *The Zoroastrian Faith: Tradition and Modern Research*. Montreal: McGill-Queen's University Press, 1993.

North, John David. *Horoscopes and History*. London: Warburg Institute, University of London, 1986.

Olson, Donald W. "Who First Saw the Zodiacal Light?" *Sky & Telescope* 77, 2 (Feb 1989): 146-148.

O'Reilly, Bill. Dugard, Martin. *Killing Jesus: A History*. New York: Henry Holt and Company, 2013.

Paschalis, Michael. *Virgil's Aeneid: Semantic Relations and Proper Names*. Oxford: Oxford University press, 1997.

Pearce, Jonathan. *The Nativity: A Critical Examination*. Fareham: Onus Books, 2012.

Pingree, David. "Astronomy and Astrology in India and Iran", *Isis* 54, 2 (1963): 229-246.

— "Historical Horoscopes", *Journal of the American Oriental Society* 82, 4 (Oct-Dec 1962): 487-502.

Phil Plait, "Starry, Starry Night", in *There's Probably No God: The Atheist's Guide to Christmas*, ed. Adriane Smith (London: Friday Project, 2009), pp. 59-67.

Poole, Michael. *The 'New' Atheism: 10 Arguments that Don't Hold Water*. Oxford: Lion Hudson, 2009.
Price, Robert M. *Night of the Living Savior*. Cranford, NJ: American Atheist Press, 2011.
Ramsey, John T. "Mithridates, the Banner of Ch'ih-yu, and the Comet Coin", *Harvard Studies in Classical Philology* 99 (1999): 197-253.
Ramsey, John T. Licht, A. Lewis. *The Comet of 44 B.C. and Caesar's Funeral Games*. Atlanta: Scholars Press, 1997.
Ratzinger, Joseph. *Jesus of Nazareth: The Infancy Narratives*. New York: Image Books, 2012.
Roberts, Courtney. *The Star of the Magi: The Mystery that Heralded the Coming of Christ*. Franklin Lakes, NJ: New Page Books, 2007.
Rochberg, Francesca. *Babylonian Horoscopes*. Transactions of the American Philosophical Society vol. 88, pt. 1. Philadelphia: American Philosophical Society, 1998.
Rosenberg, Roy A. "The 'Star of the Messiah' Reconsidered", *Biblica* 53, 1 (1972): 105-109.
Sagan, Carl. Druyan, Ann. *Comet*. New York: Random House, 1985.
Schoedel, William R. *Ignatius of Antioch: A Commentary on the Letters of Ignatius of Antioch*. Philadelphia: Fortress Press, 1985.
Schonfield, Hugh J. *The Original New Testament*. London: Firethon, 1985.
Segal, Robert A. *In Quest of the Hero*. Princeton: Princeton University Press, 1990.
Sigismondi, Costantino. "Mira Ceti and the Star of Bethlehem", *Quodlibet: Online Journal of Christian Theology and Philosophy* 4, 1 (Winter 2002).
Sim, David. "The Magi: Gentiles or Jews?" *Hervormde Teologiese Studies* 55, 4 (1999): 990-996.
— "Reconstructing the Social and Religious Milieu of Matthew: Methods, Sources, and Possible Results", in *Matthew, James, and Didache*, eds. Huub van de Sandt, Jürgen Zangenburg (Atlanta: Society of Biblical Literature, 2008), pp. 13-32.
Sinnott, Robert. "Thoughts on the Star of Bethlehem", *Sky & Telescope* 36 (1968): 384–386.

Stasiuk, Garry T. "The Star of Bethlehem Reconsidered: A Mythological Approach." *Planetarian* 10, 1 (1980): 16–17.
Stern, Sacha. *Calendars in Antiquity: Empires, States, and Societies.* Oxford: Oxford University Press, 2012.
Strauss, David Friedrich. *Das Leben Jesu kritisch bearbeitet.* Tübingen: C.F. Osiander, 1835.
Teres, Gustav. *The Bible and Astronomy: The Magi and the Star in the Gospel.* Oslo: Solum Forlag, 2002.
Tester, S. J. *A History of Western Astrology.* Woodbridge, Suffolk: Boydell Press, 1987.
Thomas, Paul. *Flying Saucers through the Ages.* trans. Gavin Gibbons. London: Neville Spearman, 1965.
Thompson, R. Campbell. *Reports of the Magicians and Astrologers of Nineveh and Babylon in the British Museum.* London: Luzac, 1900.
Thorndike, Lynn. *History of Magic and Experimental Science.* New York: The Macmillan Company, 1923-58.
Tipler, Frank. "The Star of Bethlehem: a Type Ia/Ic Supernova in the Andromeda Galaxy?" *The Observatory* 125 (2005): 168-174.
— *The Physics of Christianity.* New York: Doubleday, 2007.
Toomer, G. J. *Ptolemy's Almagest.* Princeton: Princeton University Press, 1998.
Tschudin, Max. "Das Horoskop von Jesus-Christus—ein Versuch", *Astrologie Heute* 52 (Dec/Jan 1994/5): 8-11.
Upham, Francis William. *Star of our Lord: or, Christ Jesus, King of all worlds, both of time or space. With thoughts on inspiration, and the astronomic doubt as to Christianity.* New York: Nelson & Phillips, 1873.
Van Tassel, George. *When Stars Look Down.* Los Angeles: Kruckberg Press, 1976.
Vines, Michael E. *The Problem of Markan Genre: The Gospel of Mark and the Jewish Novel.* Leiden: Brill, 2002.
Voigt, Heinrich. *Die Geschichte Jesu und die Astrologie: Eine religionsgeschichtliche und chronologische Untersuchung zu der Erzählung von den Weisen aus dem Morgenlande.* Leipzig: Hinrich, 1911.
Wallace-Hadrill, Andrew. *Suetonius: The Scholar and his Caesars.* New Haven: Yale University Press, 1984.

Warmington, B. H. *Suetonius: Nero*. Bristol: Bristol Classical Press, 1977.

Weiss, J.-C. "2000 Jahre Jesus Christ", *Astrologie Heute* 52 (Dec/Jan 1994/5): 12-16.

Yarshater, Ehsan. *Encyclopaedia Iranica*. London: Routledge, 1982-2004.

Zaehner, R. C. *Zurvan: A Zoroastrian Dilemma*. Oxford: Clarendeon Press, 1955.

— *The Dawn and Twilight of Zoroastrianism*. New York: Putnam, 1961.

Zerubavel, Eviatar. *The Seven Day Circle: The History and Meaning of the Week*. New York: Free Press, 1985.

Milton Keynes UK
Ingram Content Group UK Ltd.
UKHW010302010624
443378UK00001B/38

HOSTILE
FOR THE
Holidays

ERIN HAWKINS

Copyright © 2024 by Erin Hawkins

ISBN: 979-8-9897817-5-1

All rights reserved.

No part of this book may be reproduced in any form or by any electronic or mechanical means, including information storage and retrieval systems, without written permission from the author, except for the use of brief quotations in a book review. To obtain permission to excerpt portions of the text, please contact the author at authorerinhawkins@gmail.com

This is a work of fiction, created without the use of AI (artificial intelligence) technology. Names, characters, businesses, places, events and incidents are either the products of the author's imagination or used in a fictitious manner. Any resemblance to actual persons, living or dead, or actual events is purely coincidental.

Any use of this publication to "train" generative AI technology to generate text is expressly prohibited.

Cover design by ShaynaBCreative.com

Edited by Chelly Peeler inkedoutediting.com

ALSO BY ERIN HAWKINS

Reluctantly Yours
Unexpectedly Mine
Accidentally Ours
Surprisingly Us

Best Laid Plans
Not in the Plans

AUTHOR'S NOTE

Please be advised Hostile for the Holidays is a spicy romantic comedy with **open-door** romance, and **on-page** sexual content and profanity. Mature readers only.

*For everyone who needs a warm hug this holiday season—
this book was mine, and I hope it can be yours too.*

ONE
STELLA

I DON'T KNOW why people complain about traveling during the holidays. I love the hustle and bustle of the crowds, the anticipation of fellow travelers who are excited to reunite with family and friends, and of course, the festive décor and music in the terminal. It's the most wonderful time of the year after all.

For me, everything has gone smoothly this afternoon.

Traffic to LaGuardia was light, I checked my bag with plenty of time. I was even able to snag the last Cobb salad at the Grab N' Go station for a light lunch. Now, I'm at the gate and ready to board, so I can relax.

My eyes scan over the waiting area. There's a couple whose two kids are playing a game of cards across the open seat in one of the airport seating banks. An elderly couple looking at a phone screen together and talking. It's calm. It's chill.

Everything is going according to—oh my god. My eyes snag on a familiar face waiting in the boarding line, then I do a double-take because it's not possible. He's not supposed to be *here*.

But sure enough, it's *him*.

Thick, wavy copper-brown hair attached to a face with obnoxiously perfect bone structure. I mean, whose jaw is that sculpted? And why are his cheekbones perfectly proportioned to his sharp nose and full lips? Also, the dark-framed glasses he's wearing aren't giving nerdy tech guy at all.

Glancing down at his phone, he moves forward with the line. He's wearing dark jeans and a crocheted red holiday sweater with a snowflake on it that I wish looked silly on him but unfortunately the color compliments his warm skin tone. And the worst of it is he looks good. Better than good. Downright delectable.

I make an effort to clear my dry throat.

He shifts the wool coat from one arm to another and tucks his phone into his back pocket.

A moment later his head swings in my direction, and I instinctively slip behind a large round column next to the trash receptacles.

I think he saw me. We only locked eyes for a half a millisecond, but that's all the time I need to confirm that it is, without a doubt, him. I'd know those hazel eyes glinting with intensity anywhere.

Jasper Jensen, my childhood rival, and nemesis, is boarding my plane.

What's he doing in New York? He's supposed to be hidden away in his gated Silicon Valley mansion with all the other tech nerds creating ground-breaking technological advances.

I don't keep tabs on Jasper, but I'd have to be living under a rock to not know about his successes.

Jasper is the CEO of his own company, Jensen Innovations. It's some cutting-edge VR/AR technology that is used for corporate training and education. That's right. He's a tech billionaire and he's gorgeous. And he's making that god awful holiday sweater with a large snowflake on it look good. How many guys can say that?

With my back pressed to the hard concrete post, I take stock of my body.

Beneath my cashmere cardigan sweater my heart is racing a mile a minute. My palms are sweaty, making it difficult to keep a grasp on my leather travel bag, and beneath the waistband of my designer jeans, my tummy is tingling with an edgy, anticipatory buzz.

All the ease I was feeling earlier has been chased away by Jasper's sudden appearance.

It's worse than seeing an ex. Coming face to face with my childhood rival is like going to battle in *The Hunger Games* and I was not prepared for that today. I've got nothing in my arsenal. No witty comebacks or one-upping stories. In the rush to get out of my apartment earlier, I don't even think I put on deodorant.

The unnerving thing about this moment? I have all the boxes of a successful life checked. I was just promoted to creative director at the lifestyle brand I work for, East & Ivy. I'm the youngest creative director in the industry right now, and my branding ideas have sent sales and advertising skyrocketing. I have an apartment in the trendy and vibrant Chelsea neighborhood, and my social calendar is filled to the brim.

Or at least it's filled with first dates that amount to nothing more because finding a man to date in New York City is like trying to find a lost sock at the laundromat. It's an impossible feat. I'm never going to find the match.

And while logically I know I have time to meet the man of my dreams, there's something about the fact that my younger sister is getting married in less than two weeks that has caused my brain to hyper fixate on the fact that I'm still very single, and nowhere close to finding the one.

But other than my relationship status, I'm living my best life.

Because I'm an overachiever. I always give everything one hundred and ten percent effort.

It's the reason for Jasper and my rivalry. Since that fateful day in second grade when he told me boys were smarter than girls, a modern-age battle of the sexes began. Nothing was off limits. We competed for top grades, top honors, and generally aimed to outdo each other in everything we did. In fifth grade when we selected instruments for band, I desperately wanted to play the clarinet but I was sick the day instruments were assigned and I got stuck with Jasper on the drums. To my parents' dismay, I practiced day in and day out to hone my rhythmic skills on the drum line.

Then, we both made drum major our senior year in band, but Jasper and I had differing opinions on how things should be run, so poor Mrs. Jones, the band director, had to break up our arguments more than a handful of times. There was the famous half-time show where half of the band followed my direction, while the other half went along with Jasper. It was complete chaos and Mrs. Jones required us to take turns each game directing the band to avoid another mishap like that.

But I'm an adult now. I can choose to not let Jasper Jensen get under my skin. I'm going to march right over to the line and get on this plane without saying a word to him. Without giving him the satisfaction of knowing that seeing him today has rattled me.

But first, I'm going to peek around the post to make sure the coast is clear.

For a moment, I think about taking another flight, but that won't work.

My sister, Sadie, would kill me if I didn't get to Cedar Hollow today. She's been texting me for the past two days about all the wedding stuff that needs to be done. How stressful it is going to be to fit Christmas and her wedding into the next ten days.

I resisted the urge to tell her that Christmas was already scheduled, so it's not like it came out of nowhere when she picked New Year's Eve as her and Tom's wedding date.

Knowing I can't avoid getting on the plane, I set my leather travel bag on my carry-on and stealthily make my way to the gate. As I board the plane, I put on my oversized sunglasses hoping they will act as an invisibility cloak and make it impossible to recognize me.

I'm certain Jasper will be in first class, so I just need to make it past that section of the plane and then I'll be in the clear. I pull the latest issue of *Vanity Fair* from my bag and discreetly hold it in front of my face. I'm moving quickly through the aisle, refusing to make eye contact with anyone, and just when I've cleared first class, I slam into the back of the person in front of me, dropping my magazine.

"I'm sorry," I apologize as I reach down to grab the magazine. As I stand, my eyes lock on the man in the red sweater in front of me. In my attempt to avoid Jasper, I ran straight into him.

What the hell? Why isn't he seated in a cushy oversized chair and being served a glass of champagne?

His full lips offer up a sly grin. "Stella St. James, as I live and breathe."

I swallow hard at the sound of his deep baritone, but refuse to let the husky charm of his vocal cords affect me. Or that perfectly arranged smirk that is somehow both enigmatic and friendly at the same time.

But there's nothing friendly between me and Jasper.

"Jasper Jensen, please die and decay," I mutter.

"Glad to see you're still keeping things interesting." He laughs, nodding at the sunglasses I'm wearing. His laugh is the sound of my childhood and teenage years, only a few octaves lower now that he's a full-grown man.

I ignore him, focusing on the passengers in front of us loading their carry-ons into the overhead bins and taking their seats, but his presence puts me on edge. I'd expected him to stop in first class, you know, the whole billionaire thing, and now I'm wondering why he's even on a commercial flight. He must have a company jet.

My curiosity overrides my instinct to not engage with him.

"What are you doing on my plane?" I ask, my tone accusatory.

"This is your plane?" He turns around to smirk at me. "Stella Skyways? I had no idea."

"You know what I mean. Why are you in New York? I thought you lived in LA."

"Keeping tabs on me, Stell?" His smile is so confident, I want to smack it right off his face.

The shortening of my name sends a rush of adrenaline through my veins.

Stell from Hell is the nickname that Jasper gave me in middle school. He used it whenever he thought I was being dramatic or over the top, which in his opinion was often. In turn, I called him *Jasper the Disaster*, but it didn't have the same effect of riling him up like his nickname for me did, because truth was, he wasn't a disaster. He was a straight-A student, all-star athlete, and popular guy who excelled at everything he did.

"You wish." I barely refrain from sticking out my tongue, because it doesn't matter that I'm twenty-eight years old, Jasper makes me feel like I'm seven and I've got to hold my own on the playground.

"I was here for a business meeting, and now I'm flying home for the holidays. Is that okay with you, Stell?"

I wait for him to add *from hell*, but it doesn't come.

We move forward down the aisle, and I wait for him to stop at his seat so I can pass.

He doesn't stop. He keeps moving. And as we get closer to the back of the plane, my stress level is peaking.

With his proximity my initial flight instincts shift into full-on fight mode.

"By the way, your sweater is ugly."

"Thanks." He turns back to smile at me. "My ninety-year-old grandmother crocheted it for me."

I reconsider my comment because that's really sweet but Jasper doesn't deserve an ounce of kindness from me. He made sure of that our senior year in high school when he spread a rumor that I was in training to become a nun and going to join a convent after graduation. I didn't find out about it until I was dateless for prom and asked Jamal Lancaster to go with me. He declined because he said he actually wanted to have fun and not be going with a nun.

"This has been a super fun reunion, but I'd like to get to my seat so if you could move it along."

"My seat is right there." He points in front of me to row thirty-three where next to an older gentleman, there are two unoccupied seats.

All the color drains from my face.

"You've got to be kidding me."

He shows me his phone. Thirty-three D. He has the seat next to mine.

How is that possible? Of all the airports in the city, of all the flights and seat assignments, how the hell did I end up next to Jasper Jensen on this four-plus-hour flight?

A flight attendant wearing a reindeer antler headband with red and green jingle bells and a cheery smile approaches us. "Please find your seat so passengers behind you can pass by."

Jasper motions for me to claim my middle seat but I'm not ready to give in that easily.

I ignore him and turn my attention to the flight attendant. "About the seat situation, is there a chance I could move to an empty seat?"

"I believe it's a completely full flight, so I'm afraid all the seats are taken."

"Maybe I could switch with someone? Anyone?" I plead, looking around me, but nobody wants a middle seat toward the back of the plane.

She gives me a placating smile, then motions for me to take my seat.

"Fine," I growl, defeated. I don't plan to talk to him anyways. I'll put on my headphones and listen to one of my friend Pippa's romance novels on audiobook.

Jasper has already placed his carry-on overhead, and is waiting in the aisle for me. I attempt to shove my carry-on suitcase under the seat, but the flight attendant stops me.

"Miss, that will have to go overhead."

"Sure. No problem." That's what I say, but I know lifting this overloaded carry-on is going to be a challenge. Lift with the legs, engage the core. The suitcase doesn't make it past my knees. It's clear from this demonstration of my weak muscles that I need to start working out.

"If it's oversized, we'll need to check it," the flight attendant warns me, the bells on her headband jingling ominously with the shake of her head.

"I—" I start to protest because this suitcase is important. It's got all the new design concepts I've been working on as well as my sketch book, presents for my family, and my bridesmaid dress.

"We'll make it fit." That's Jasper as he easily lifts it above our heads and into place in the overhead bin. I'm a statue as his body curves over mine, his broad chest and that gawdy sweater of his brushing against my back. I get a whiff of his cologne and damn it, he smells great, too. Meanwhile, my deodorant-less armpits are dripping sweat.

Jasper winks at the flight attendant and she blushes.

"Get a room," I hiss when she's gone, then take the middle seat.

"I was being nice," he says, dropping into the seat next to mine.

"You were flirting while she's trying to work."

"Do you even know what flirting looks like?" he asks.

"Excuse me?" I scoff. The comment feels like a direct dig at my relationship status. That being single and alone and most likely to be eaten by my cats if I had cats. Which is why I will not be adopting any cats, for any reason, ever. "I have plenty of experience with flirting and men flirting with me."

"Sure." He nods. He's agreeing with me but it's his tone that gets to me. It's the one that tells me he's placating me, which from Jasper is like putting a match to gasoline. It's what Jasper does. He riles me up then brushes off a conversation that I was winning as not a big deal. Well, it's a big deal to me.

"For your information, I am a prolific dater. I have been

on plenty of dates with highly desirable suitors." I push my leather travel bag under the seat in front of me and buckle my seatbelt.

"Yeah?" His brows lift, and it's unfair how attractive he looks making that quizzical face. "Are you currently in a relationship?"

"Not that it's any of your business, but no. I haven't found the right guy to commit to."

"And these dates you've been on, how many of them were second dates?" he asks.

My hackles rise, wondering what he's inferring.

"What does that have to do with anything?" I snap.

"Can't start a relationship without a second date." He smirks.

I think back to the dates I've been on this year and other than the doctor that had to leave in the middle of our date so we rescheduled, there haven't been any second dates.

I find Jasper staring at me, a peculiar look on his face that has me going on the defensive.

"That doesn't mean anything. And you're one to talk. Where's your girlfriend?" I motion to the empty space around us to prove my point.

"Don't have one."

"Ha!" I exclaim, shooting my finger in the air like I've solved an age-old mystery.

The woosh and subsequent shifting of the plane lifting off has me reaching for the arm rest. However, it's not the plastic arm rest my fingers grip, but Jasper's warm, muscular forearm.

I hadn't even realized we were taking off.

"Excuse me."

"You're excused," Jasper says, leaning back into his seat.

"That's not what I was referring to. You need to move

your arm." I nudge at his forearm with my elbow. "Everyone knows the courtesy is to give the person in the middle seat two arm rests."

"Nobody knows that. You're making it up."

"Well, they should. It's the way to balance out the discomfort of being the middle seater."

"Aren't you going to tell him about this policy?" Jasper lifts his chin toward our row companion.

The man's large frame is sprawled out, his forearm taking up the entirety of the arm rest between us. His mouth, which is covered by a thick brown mustache, is gaping open because he's already asleep.

I open my mouth to argue with Jasper, but that's what he wants so I snap it closed and decide to ignore him for the rest of the flight.

Leaning forward, I reach into my travel bag for my noise cancelling headphones. They were a gift from Sadie last Christmas and I'm thankful today more than ever for them.

Jasper makes me cranky and moody. And my stomach hurt.

That's a new one.

Ignoring my body's reaction to Jasper, I power on my Bluetooth headphones, then start playing my audiobook where I left off last night. It was right in the middle of a tension-filled sex scene. That's exactly what I need to distract myself from Jasper's presence. And I love the fact that I can listen to this sexy audiobook and he has no idea. It's my little secret.

I wait for the audiobook to start, but nothing happens. I reach down to hit play again, but it says it's already playing. I can see the seconds ticking down on my screen. That's odd.

A hand lifts the side of my headphones. Why can't he leave me in peace? Also, why is this audiobook not playing?

"What?" I ask sharply.

"Your audiobook is playing out loud."

I yank my headphones off and immediately hear the male narrator describing in explicit detail how he's going to make the female character come. My hands fumble for my phone and try to hit the pause button, but I accidentally swipe the volume and the groan of the narrator's voice gets louder.

Beside me, Jasper takes my phone and hits pause like a person with steady fingers can.

I glance up to see the woman in front of me giving me a scolding glance over her seat.

"Sorry," I whisper, wanting to sink underneath my seat.

"I'd ask what you were reading, but I think the entire plane heard it."

"Oh, zip it." I put the malfunctioning headphones back in my backpack. "I don't care what you think."

"I didn't say it wasn't intriguing. And now I'm on the edge of my seat wondering if Wyatt is going to make Rosie come."

I stare at him in shock. This isn't happening. I'm not going to discuss my spicy romance audiobook with Jasper. There's no way in hell.

My stomach gurgles.

Then hisses. And then there's an unsettled roil.

With one hand on my stomach, I pause.

I can't be hungry. I ate that salad only an hour ago.

I'm not prone to motion sickness and there hasn't been any turbulence. It's probably my stomach revolting at being near Jasper. A physical response to being in the presence of my arch nemesis.

Then a familiar wave washes over me.

Not the orgasmic kind. That kind of wave is typically self-induced by my vibrator after listening to a particularly spicy scene.

It's nausea. My skin goes clammy and my body sways as I reach to the seat in front of me for support.

Oh, no.

"Jasper, I need to get up."

"What? Why?" His brows pinch down, meeting the bridge of his glasses in concern.

But he doesn't move.

"I need you to move. Now!"

He's going at a sloth's pace. Before he even has his seatbelt unbuckled, I'm climbing over his lap. Once I'm in the aisle, I make my way toward the lavatories, but both signs are illuminated in red with an X over them.

I hover outside hoping someone will come out soon, but I quickly realize I'm not going to make it.

I clamp a hand over my mouth and with maximum effort to hold everything in, I reverse direction. The lavatories up ahead are green, but the flight attendants have the drink cart blocking the aisle.

Oh, no. The contents of my stomach are bearing down on me.

I can't throw up next to strangers.

I have to get back to my seat.

I need one of those bags.

A vomit bag.

Oh, god.

When Jasper sees me coming, he smirks.

"Back already?" he asks in a mocking tone.

Panic grips me.

"I need a bag!" I yell, but Jasper's eyes only narrow in

confusion. It's a moment later when he sees me holding a hand over my mouth that he springs into action, reaching forward into the seatback pocket for the airplane-provided sick bag.

But he's not fast enough. And I know I'm not going to make it.

I reach for the hem of my sweater to make a makeshift bowl or bag or anything that is going to catch what is coming up.

Then, I let it all go.

TWO
JASPER

I'VE IMAGINED a moment like this before. Stella St. James' head cradled in my lap as she looks up at me with sleepy eyes, her perfect lips curved into a content smile.

"Your lap is ridiculously comfortable." She sighs. "I hate it."

Maybe not this exact moment, but it's a start because I've been in love with Stella St. James for ten Christmases, and this year, I'm determined to put our childhood rivalry behind us and tell her how I feel.

To say things are not off to a good start is all about perspective.

I could have done without her being violently ill for the past two hours, but things are looking up now. She drank some water and has been able to keep it down for the past thirty minutes.

Once Stella threw up for the third time, the man next to us was relocated to another seat, so she's now stretched out across two seats with her head in my lap. I like her like this. Not sick, but softened enough to let her guard down.

She shifts, adjusting her head to get more comfortable. "I should mention this is a revenge plot and you fell for it."

Okay, maybe her guard isn't completely down.

"I thought to myself what is the grossest, most repugnant thing I could do to Jasper, and throwing up my Cobb salad was the first thing that came to mind."

I have to smile. Even in the throes of food poisoning she's thinking about me.

"I'm flattered you went to such lengths to make my flight miserable, Stell. A less committed rival would have simply opened their water bottle mid-flight and sprayed me in the face with it."

"There's still time for that," she says softly.

When I helped the flight attendant clean up after Stella got sick in the aisle of the plane, she'd commended me on being such a good boyfriend. I didn't bother to correct her. Partly because I liked the label but also because it occurred to me that what person, besides a significant other, would clean partly digested food and stomach acid up for someone? It was disgusting but I'd do it again in a heartbeat for Stella.

"It was the last salad, Jasper. I thought it was a sign of good luck, but I should have known it was a bad omen. And now the universe has put me right in the arms of the enemy and I'm too weak to fight back."

"Shh. Or I'll start playing your spicy romance novel on your phone again."

"You enjoyed it. Don't lie."

I chuckle because, yeah, I did. It had been fascinating to hear what Stella was playing on her headphones. A glimpse into something she enjoys in her current life. One that I know so little about living on the other side of the country.

I brush a piece of her hair off her face and her nose

scrunches, but from my observations of Stella's disgusted nose scrunch over the years, it's a half-effort at best.

"I know what you're doing. You're trying to look like the good guy. Make everyone on the plane think you're sweet and thoughtful and unreasonably handsome. But they don't know you like I do, Jasper."

"Unreasonably handsome?" The corner of my mouth pulls up.

Her eyes open suddenly, and for a moment I enjoy the way those blue gems stare up at me. For the last hour or so she's been mumbling all sorts of things to me. It's like having an inside track into her mind, I never thought I'd have such an opportunity. Over our teen years, Stella had made sure to keep a wall up between us. For once, I'm able to look over the top and see what treasures are hidden on the other side.

She lifts a hand toward my face, her index finger tracing lazily along the slope of my nose.

"Don't argue with me, you'll never win," she says before turning her head away.

She shifts her head again, like she's trying to get comfortable, and I realize her earrings are likely the issue.

"Let's take these out." I run a thumb over the eye-catching earrings, watching the green and red crystals shimmer under the airplane cabin lights.

"Fine. But keep them safe. They're my favorite holiday earrings and since my sweater is already ruined, I don't want to lose them, too."

She's not wrong. The soft cardigan sweater I'd brushed up against when we were fighting over the arm rest didn't fare well. Luckily, she had a tank top on underneath. The stretchy cropped tank top is low cut and has been giving me an eyeful of Stella's cleavage for the past two hours.

Pulling my gaze from her breasts, I will myself to focus on the task at hand.

I gently remove the wreath-shaped baubles and place their backings on before tucking them away in Stella's leather travel backpack.

With her earrings out, she snuggles into my lap, looking far more relaxed.

"I can't keep my eyes open. I have to rest," she mumbles. "Please don't write on my face. At least not with a permanent marker. I've got my sister's wedding next week and she'll kill me if I've got Sharpie on my face in her wedding photos."

I brush my thumb against her cheek, grateful for any excuse to touch her.

"I could draw a little mustache." My finger traces a swirl above her plush lips on one side, then the other. "Maybe some devil horns." I sweep my fingertip along her forehead, sketching the invisible horns.

"Come on. That's so unoriginal. That's what you did to me in the class photo in third grade."

"It's a classic."

I'm quiet a moment, my fingertip continuing to trace lines around her face.

"Are you excited for Sadie's wedding?" I ask.

She makes a noncommittal sound. Maybe that's all she has energy for right now or it could also be that she's not looking forward to it.

"Do you have a date?" I ask.

"I'm not answering that."

"It's a simple yes or no question."

I've remained calm for this entire plane ride, but the moment we start talking about Stella's potential date, my heart starts to race. She said she was single, not in a committed relationship, but she could still have a casual date to her sister's wedding.

"Not from you." She takes in a deep breath, then exhales.

"I'm trying to make conversation." From her silence, I can tell she's not in the mood for it, so I deflect to taking care of her. "You ready to have some more water?"

"Yeah."

I prop her up enough so the water doesn't spill down her front, then hold the plastic cup so she can drink.

"If you tell anyone about this, I will deny it, then make your life miserable."

"More than you already do?" I quip, just for old times' sake.

"Funny. Just remember, if you thought I was a stubborn menace in high school, think of the resources I have now. A credit card, for starters. Friends who can troll you on social media. The possibilities are endless."

She's adorable.

"Sure, Stella. I won't tell anyone."

She shivers.

"Are you cold?"

"No."

Her teeth chatter while goosebumps break out across her skin. Not to mention the two hard peaks poking through the front of her tank top. Fuck.

She's defiant at best, a liar at worst.

I ease her to sitting, so I can stand from my seat and reach the overhead compartment.

"What are you doing?" she grumbles. "I was comfortable."

"I'm grabbing your suitcase to find you another shirt."

We'd disposed of Stella's sweater. She cried and

claimed it was her favorite as I double-bagged it in hazard material bags and stored it under the window seat.

"I don't have another shirt in my carry-on. It's only presents and my bridesmaid dress. And I can't wear that. Sadie will kill me if I get it dirty."

Instead of reaching for Stella's carry-on bag, I pull out mine.

"Here." I hand her a sweater. "Since I know you're such a fan."

She stares at it, part awe, part disgust. It's identical to the one I'm wearing, red with a snowflake on it, but in a smaller size.

"It's Juniper's, so I'll need it back."

She nods, then pulls it over her head, before we resume our position. Stella's head resting peacefully on my lap while she sleeps for the remainder of the flight.

THREE
STELLA

JASPER CLEANED UP MY VOMIT.

That knowledge makes me both want to laugh and cry.

If I wasn't so self-conscious about my current condition, I would give myself props for this unintentional revenge plot, because if Jasper thought he got the best of me by being my seat mate, then I showed him.

I could have sworn I heard the flight attendant fawning over how he was such a good boyfriend. Pfft. She knows not what she speaks of. And I would never. *As if.*

Things happened on that flight that I'm not proud of, and throwing up isn't even one of them.

I snuggled my head in Jasper's lap. Then, I vaguely remember touching his nose and I might have called him handsome. It's all part of the delirium I experienced once the worst of the food poisoning had passed, when dehydration and exhaustion had taken over.

As I make my way off the airplane, wearing Juniper's sweater, I throw the hazardous waste bag containing my beautiful cashmere cardigan into a nearby trash can and say a prayer for it. Even if it could be cleaned, the memories of what happened are too haunting to move forward in our relationship.

By the end of the flight, my stomach started feeling better, and as my fatigue and dizziness lessened, I felt the need to put distance between me and Jasper. He's seen me at my worst now, and I know it's only a matter of time before he uses it against me in some way. Needing to pull myself together so I can cling to what shred of dignity I have left, I take advantage of the fact that Jasper is helping an elderly couple get their suitcases down from the overhead bins, and rush off the plane without saying goodbye.

It's fine. I'll dry clean Juniper's sweater, then set it on his family's porch and we'll never speak of this again.

All I want to do now is get my suitcase from baggage claim and run into the comforting arms of my family.

Nothing as extravagant as the montage of airport arrivals at the end of *Love, Actually*, but my parents and Sadie gathered behind a simple 'welcome home' banner would suffice. It has been months since I saw them.

When I come up the escalators to the main terminal, there's a crowd of people waiting but no one there for me.

No biggie.

On my way to baggage claim, I pull out my phone to check my messages, to see where I should meet them, but there aren't any.

That's odd.

I'm scrolling through my recent contacts, a moment away from calling my dad, but then I make eye contact with a guy across the baggage claim carousel.

No. No. No. This can't be happening.

For the second time today, I want to hide.

I pray that I can grab my suitcase and be on my way

before he sees me, but that's when I see the sign he's holding. And it's in the next second that his eyes latch onto me.

There, smiling and waving a sign with my name on it in colorful gemstones, is my sisters' fiancé's cousin, Daniel.

Fuck.

I didn't think this day could get any worse, but in fact it can.

Daniel and I spent one drunken night together after a Noah Kahan concert last summer. It was an okay time—I mean with Daniel, because Noah's performance was phenomenal—but nothing I wanted to pursue and while I've tried to keep things cordial because he's going to be part of my sister's family once she marries Tom, I finally had to tell him that there's nothing between us. That didn't help. He only came on stronger thinking I was playing hard to get. Over the past few months, he's been sending me flirty text messages which I've ignored hoping he will get bored and move on.

While I knew he'd be at the wedding, I didn't anticipate seeing him at the airport. And him standing by baggage claim with my name on a sign was not on my 'home for the holidays' bingo card.

And from the way he lights up when he sees me, it appears he hasn't moved on.

"There you are, gorgeous."

He moves in for a hug, but I hold my travel bag between us to keep from our bodies pressing together so it ends up being an awkward side hug.

"Wow, Stella. Red is your color." He whistles loudly. An annoying catcall whistle that makes me cringe. "You look amazing."

"Thanks, I have vomit in my hair," I say flatly. "What's

with the sign?" I glance around, looking for any member of my family to swoop in and save me.

"I'm your ride." He opens his arms wide and his potent cologne hits my nose.

Even if I wasn't recovering from food poisoning, I don't think I'd be able to stomach the overpowering scent.

"I thought my dad was coming to pick me up?" I ask, dread making my stomach churn again.

"I volunteered. Told your dad it would give us time to catch up." He winks and it's awkward. Nothing like the smooth charm Jasper had displayed on the plane. Oh god, now I'm comparing other guys to Jasper. I think the plane flew through a wormhole and landed me in an alternate universe.

"But why are you here so early? I thought guests weren't coming until after Christmas."

"Tom's here and you know he's like my brother. We spend the holidays together now that my parents are overseas, so your mom offered for me to stay in your family's guest room." A huge smile pulls across his face. "Isn't that great? We'll have plenty of time to hang out. Rekindle that spark between us." He gives me another twitchy wink and I think I might be sick again.

Daniel's annoying and a bit pushy, but he's nothing I can't handle.

The question is, do I want to spend my two-week holiday vacation heading off his advances? His over-the-top flirting and barrage of innuendos. Visions of my peaceful, relaxing time at home are now filled with the scent of Daniel's cologne and the sight of his leering smile directed at me.

Then there's the fact that I don't have a date to Sadie and Tom's wedding. I'm a bridesmaid and Daniel's a groomsman so we'll be together at every event with no one to run interference between us and make Daniel understand once and for all, I'm not interested.

I need a moment to process my thoughts.

"I've got to—" I motion, turning to look for the nearest women's restroom, but I run smack into a man's chest.

Jasper.

When my eyes lift to his, it's the oddest moment. Hours ago, I was avoiding him, now the sight of his hazel eyes and knowing smile are like a lifeline.

It must be because in this moment when I'm faced with Daniel, a past romantic mistake that I'm going to have to endure through the holidays, Jasper seems like the least unpleasant choice. I'd rather bicker with Jasper than be the subject of Daniel's affection. The thought has my mind spiraling. I never imagined Jasper Jensen would be a safety net.

And that's what I need right now.

A shield. In human form.

The only way to show Daniel I'm truly not interested is if I'm with someone else.

I need a boyfriend to keep Daniel's advances at bay.

I don't have a boyfriend.

My thoughts ping-pong as I try to come up with a solution to this mess.

But just maybe, I could pretend.

Would Jasper play along?

He was awfully good at pretending to be concerned about me on the plane.

Jasper's gaze lifts from me to Daniel, then back to me. I do my best to send him a silent plea. I hate the thought of pleading with Jasper but the alternative in this situation is worse.

"There you are," I say, leaning into Jasper's clean scent to escape Daniel's overly manufactured one.

"You left your water bottle on the plane." He hands me the pastel pink metal bottle.

"Oh, thank you so much, *babe*." Jasper's brows lift at the term of endearment. "You know it's my favorite one. I don't know what I would do without such a thoughtful *boyfriend*." I wrap my arms around Jasper's waist and am surprised by the muscles that are present underneath his sweater. How firm and corded they are.

"Boyfriend?" Daniel groans behind me. "Sadie didn't mention you were seeing anyone."

I turn to face him, trying not to delight in his defeated expression.

"It's a surprise. I haven't told my family yet." I motion between the two of them. "Daniel, this is Jasper. Jasper, Daniel. Daniel is Sadie's fiancé's cousin and he's staying with my family over the holidays." I do my best to keep my smile even but Jasper hasn't said anything yet and it's starting to make me squirm. Is he really going to play along? It doesn't seem like something he would do for me.

Daniel extends his hand to Jasper. "Stella and I have history. Romantic history, so I'm ready to step in if you mess this up." He motions between me and Jasper, but then his expression softens. "I like your sweaters."

I'd completely forgotten that Jasper and I are wearing matching sweaters. It's a nice touch to sell our supposed relationship, and for a moment I regret having said they were ugly.

"Thanks! Jasper's grandmother made them," I say, beaming with pride.

It's quiet again and awkward tension settles between the three of us.

Okay, what now?

I glance around the area, trying to remember what I was doing before Daniel blindsided me.

"Oh, my suitcase!" I rush over to collect it from the carousel, but right before I can claim it, Jasper's hand reaches around me, easily lifting it off the moving track and onto the ground.

"Thanks." I bat my lashes sweetly and Jasper scowls at me.

We've got to wrap this up before he blows my cover.

"Do you two need a ride?" Daniel asks.

"No!" I shout before Jasper can answer. "We've got our own transportation. Thanks."

"I guess I'll see you at your house, Stella." Daniel gives a final glance between me and Jasper before walking toward the exit, dropping the sign with my name on it in the trash on his way out.

"Okay, bye!" I give a quick wave as relief washes over me at Daniel's departure. But then I turn to meet Jasper's hard expression head on and wonder if I dodged one bullet just to be struck by another.

FOUR
JASPER

WITH DANIEL GONE, Stella turns around to face me.

On the plane, I had pulled her blonde waves into a messy ponytail. It had been a necessity when she was throwing up and now that I have a better view of her, it appears I didn't do that bad of a job. A few loose hairs frame her heart-shaped face. She's got more color now in her cheeks and lips than she did earlier, they appear to be returning to their natural rosy appearance, and her blue eyes are two bright gems piercing my soul.

She's stunning and from the way those blue eyes are slowly narrowing at me with contempt, she must be feeling better.

I'm not one hundred percent clear on what the situation was with this Daniel guy, but I know I didn't like it.

"What the hell was that?" I ask, an edge to my voice that I've never used.

That's because I've never seen Stella St. James vulnerable like she was a moment ago and it awoke something inside me.

Protectiveness.

And when Daniel said he'd been romantically involved with Stella, I'd wanted to put him in a headlock and wrestle him to the ground. At the very least, tear up his sign with her name on it because I didn't mastermind being next to her on our flight home just to be ambushed by some guy with a bedazzled sign. I'm the one that held her on that airplane when she was sick and I'll be damned if I'm going to let some guy swoop in and make a move on my girl.

The problem is, she's not my girl. She's the furthest thing from it.

Stella and I have been at each other's throats since we could form proper sentences. I don't even remember how our rivalry started, but somewhere around our teen years it spiraled out of control, taking on a life of its own. Now, it's me, Stella, and our rivalry. And three's a crowd.

I'm not upset with her, but Daniel, who was making her feel uncomfortable. But as usual, where I'm concerned, she interprets my question as a personal attack and goes on the defensive.

"Nothing. Just forget it." She grabs the handle of her suitcase and starts marching off.

"Where are you going?" I call. "Wasn't he your ride?"

She halts at the mention of how she's going to get home. That's right. We're an hour and one mountain pass from Cedar Hollow, our hometown. But then again, I wouldn't put it past Stella to make the journey fueled only by sheer determination.

Jesus, she's so fucking stubborn and I love it.

She swallows. Even on the heels of being sick, she's got that fiery passion in her.

"I thought my dad was going to pick me up, but Daniel volunteered. He's Tom's cousin and apparently staying with my family for the holidays. I couldn't handle an hour in the car with him making small talk."

"I've got a rental car. I'll drive you home."

"It's fine. I'll get a rideshare or see if the shuttle has an available seat."

"Stella, you've been sick. You're tired and dehydrated. I'm driving you home."

"Jasper—" She begins another argument but I'm losing my patience. While I like to see Stella passionate and fight for what she wants, I'm tired of fighting *with* her. I want this Christmas to be different.

"Enough with this back and forth. You can walk to the car, or I'll throw you over my shoulder and carry you out."

She blinks, like she can't believe I said that to her. But once the shock wears off, she recovers with a smirk.

"Ha! I'd like to see you try."

I lunge toward her but she squeals, then scurries away from me and toward the exit.

"That's what I thought."

I grab the handle of her suitcase, then follow her out the doors pulling both of our suitcases. Outside the airport terminal, the chilly mountain air hits us, and I watch as Stella wraps her arms around her waist and shivers.

"Where's your coat?" I ask.

"In my suitcase."

"That's a silly place for it."

"It's bulky and I didn't want it on the plane. And if I wanted your opinion, I'd ask for it."

"Good evening, Mr. Jensen." The concierge from the rental car company approaches from the curb where the Range Rover I rented is waiting. He helps us load our luggage into the back, then hands me the key.

"I should have known you weren't renting a car like normal people."

"What do you mean? This is normal."

"The rental car showing up to you instead of having to be shuttled out to the middle of nowhere to pick up your car is a bit bougie."

"No, having a driver would have been bougie." I smile. "I like to indulge in a few nice things."

Stella rolls her eyes at me. "A few nice things? You're a walking advertisement for GQ man of the year."

"I'll take that as a compliment."

"Please don't."

"Do you have a thing against successful people in general or is it my success that annoys you?"

"Do I really need to answer that?"

"You're one to talk. Creative Director at East & Ivy. Now, that's impressive."

"Stalker," she murmurs.

"It seems like you're feeling better. How's your stomach?"

"I'm fine. I don't need your concern. I'm in tiptop shape. Back to my fighting weight. A worthy opponent once again."

I flash her a soft grin.

"Glad to hear it."

"Yeah, yeah."

She buckles her seatbelt and I pull away from the airport.

"What—" I begin but Stella quickly shushes me.

"Let's play the quiet game. See who can keep unnecessary chatter to themselves the longest."

"Come on, Stell. We've got so much to catch up on.

Like Daniel, for instance. What's the deal with that guy and why did you tell him I'm your boyfriend?"

Her face pinches with annoyance. "You're so fucking nosy, Jasper." But she's got no escape route now, so I push her on it.

"Nosy? I'm trying to understand why you'd tell this guy we're dating." And how I can use this development to my advantage.

She throws her arms up.

"Fine. Daniel and I hooked up. It was a mistake. One that I'm clearly still paying for because even though I've told him I'm not interested, he thought he would show up here and we'd fall madly in love over the holidays. Or at least hook up again. So, yeah, when I saw you, I pretended you were my boyfriend so he'd back off."

Logically I know Stella has a personal life that I'm not involved in, but hearing about it and seeing a guy firsthand that she has been involved with only makes these feelings I have for her stronger. This situation with Daniel, while I hate that it makes Stella uncomfortable, is the perfect opportunity to show Stella how much I care about her.

"Do you need help, Stella?" I ask, giving her a quick glance before turning my eyes back to the road. "All you have to do is say the word."

She's quiet for a moment, arms crossed over her chest in defiance.

"And what will it cost me, Jasper?"

Nothing. I'd do anything for you.

My answer forms easily. After all, it's been circulating in my brain for years. But Stella doesn't know that, so I take a moment to think about her question.

"You need a boyfriend for the holidays." I brainstorm out loud.

"Not a boyfriend. A human shield," she asserts. "It would be fake. We wouldn't actually be dating."

"What about your sister's wedding?" I ask.

"It's on New Year's Eve. I'm sure my parents invited your family," she says begrudgingly.

"Daniel's going to be there?"

"Yeah, I told you, he's Tom's cousin. And he's staying with my family. For the next two weeks, he'll be everywhere."

"So, you'll need me to be your date to the wedding as well."

"Are you adding this all up so you can send me a bill?" she asks.

"No, my fake boyfriend services are free."

"That's shocking." Her eyes narrow. "What's in it for you?"

It occurs to me that Stella will never agree to an arrangement that is one-sided. She'll think I have an advantage over her or I'm doing her a favor which will ultimately become a future issue between us. I need a reason to have a fake girlfriend or this doesn't work.

"I'll need you to reciprocate. Be my girlfriend at my family's Christmas Eve party."

"Fake girlfriend," she emphasizes. "And why?"

"My mom is notorious for trying to set me up with women at that party. It's a tradition I'd like to avoid this year." It's a stretch of the truth. My mom tried to set me up one year and I asked her to never do it again.

We sit in silence for a few minutes while she thinks it over. It's mostly silence, there are a few mumblings and murmurings coming from Stella as she battles with her decision.

"Fine. I'll do it." She drops her head back against the head rest with a sigh. "But this changes nothing between us."

"You want to pretend to be dating, but still be at each other's throats?"

"You're right. This is going to be too hard. I'll have to tell Daniel we broke up and then find someone else to fake date." She pulls out her phone and starts scrolling. "Oh, maybe Jonah Collins. He was always nice to me."

"Jonah Collins got married last year," I say through gritted teeth, remembering him as Stella's boyfriend for six months our junior year in high school.

She scrolls again. "Max Rhodes?"

"Moved to Florida. He won't be home this year."

"Darn it." She sighs, dropping her phone into her lap. "How do you keep up with everyone?"

"I don't really. Juniper knows everything."

My twenty-two-year-old sister is attuned to what happens in our hometown. While she is finishing up her business degree at the University of Colorado in Boulder, she comes home to visit often.

Stella's face softens, a sweet smile pulling at her lips. "How is Juniper?"

"She's good. Going to graduate in the spring and wants to open a romance bookstore in town."

"Wow." Her eyes light with fascination. "That's really cool."

From the sincere look on her face, it's evident that Stella's hostility doesn't apply to my entire family, just me.

"Listen, finding another boyfriend on this short notice is ridiculous. Daniel will see right through it." I tap my fingers against the steering wheel. I need another angle. A way for

Stella to see this as a competition. "You're saying you can't pretend to be in love with me for two weeks?"

"In love with you? Ha! In this made-up scenario, we're dating, we aren't in love."

"Okay. So, Stella St. James, the phenomenal theater performer who won best actress at the 2013 Cedar Hollow High School Drama Awards, can't pull off a simple fake dating arrangement?" I follow up my challenge with a low whistle to really drive home my point.

She straightens in her seat and I know I've got her hooked. "Of course, I can."

"Then prove it."

"Hold on. What am I getting in return? I've got the acting skills, but I don't think your time as the Technical Supervisor in drama club will lend itself to this project."

She's referring to me being the lighting and sound designer for all the theater productions. It's there that I got my start learning about the technical aspects of sound and audio design, and my love for the visual arts. It became my major in college and my interest grew into the start-up that is now my billion-dollar company, Jensen Innovations.

I do have Stella to thank for that discovery. She's the only reason I joined the drama club. To be near her.

"Trust me. I've got plenty of skills that will come in handy."

She bursts out laughing. "You want me to trust you? After all the shit you've pulled?"

"Me? What about you? You're not innocent in this."

"I never claimed to be." She huffs, her voice raising with her growing frustration. "This will never work. I despise you too much."

"Come on, Stell. Is your hate for me more than your desire to avoid Daniel?"

She's quiet for a minute, her lower lip trapped between her teeth as she contemplates my question.

"Okay, Jasper. For the next two weeks, we can pretend to be dating."

A satisfied smile pulls at my lips, and for the rest of the drive, I honor Stella's request for silence while she dozes quietly in my passenger seat.

The hour drive flies by and it's not long before I'm pulling into Stella's driveway.

"Your family didn't decorate their house yet?" I ask, taking in the lack of holiday lights. It's unusual for the St. James house not to be lit up in a full holiday display. Stella used to hold a holiday light decorating contest for the block. Anyone could enter, but I know in her mind her only competition was me. My dad and I would spend hours on the weekend after Thanksgiving putting up our display. Over the years, the trees in our yard have gotten taller, and our display has gotten more elaborate, so as a gift to my parents, I hire a company to install them.

Stella clears her throat. "Um, no. My dad is probably waiting for me to help him. It's our thing."

I put the vehicle in park and get out to grab Stella's luggage. "Come on, sugar lips."

On the pavement, Stella's gaze narrows. "That's not a nickname I would ever answer to."

"How about honey pie? Sweetie? Pookie?"

She ignores the teasing nicknames and punches in the garage code.

"Goodbye, Dickface."

I make a mock discerning face. "I think you're going to need to work on that one. It's not giving loving, supportive girlfriend."

"Okay, Jass." She laughs.

A warm rush of contentment fills my chest. I like it when she calls me that.

"Why's that funny? You just shortened my name."

"It has an extra S on the end, so it's like Jasper and ass combined. J-A-S-S."

I don't bother to tell her that I like when she calls me Jas, whether it's with one S or two. And unless she's going to spell it out, no one will know the difference.

"Sure, that one will work."

I grab her suitcase and move to pull it inside the garage.

"What are you doing?" Stella wrenches the suitcase handle from me.

"I'm taking your suitcase inside."

"No way."

"How will it look if we're dating and I drop you off at the curb?" I ask.

"It's not like you just slowed down and pushed me out of the car," she insists. "Besides, I need a minute to wrap my head around all of this." She motions between us.

Our fake relationship is fledgling and if we're going to make it through the next two weeks, we'll need to take it slow.

"Okay," I relent, heading toward the car again. "Goodnight, Sparky."

Her mouth gapes open in protest. "What's that for?"

"All the chemistry between us."

She glances down at my sweater. "Goodnight, Snowflake."

"You know you're wearing the same sweater," I say, walking backwards toward my car.

She smirks. "Not voluntarily."

As the garage closes, she holds a hand to her lips like

she's going to blow me a kiss but just as the door goes down, she flips me off. It only makes me smile.

For the first time in the years since I've known her, competed with her, and been powerless to keep my heart from getting crushed by her, I'm hopeful.

FIVE
STELLA

WHEN JASPER PULLED INTO MY PARENTS' driveway, I was stunned. I even double-checked the address to make sure it's my house. It's got the same white brick and pillars, the same cardinal red door with a half-moon window at the top. The spruce trees are exactly where they've always been framing the front stoop, but there's noticeably one thing missing...Christmas decorations.

But if there's one thing I've learned in my years going toe-to-toe with Jasper, it's to never let him see my weaknesses.

That plan imploded today when I threw up on the plane, then laid in his lap like a helpless toddler. Oh, and when I asked him to fake date me to deal with the Daniel situation. That was a doozy.

I still can't believe I suggested it. And Jasper agreed. Or was it the other way around? I can't even remember how the conversation went. But it doesn't matter, I'm going to nail my part as Jasper's fake girlfriend if only to spite him.

Because that's what we do.

We out-do and one-up each other and never ever let the other one know when they get to us.

That's why there was no way in hell I could let Jasper see how my family's house not being decorated for Christmas disappointed me.

I'm holding out hope that while outdoor lights were not on the priority list this year that the inside of the house is a winter wonderland with my mom's crocheted Santas lined up on the fireplace above the stockings. Gingerbread candles burning and Christmas music on repeat in every room. The tree decked out in old school popcorn strands and ornaments Sadie and I made in the nineties that have been glued back together a handful of times.

As I move down the hallway and past the living area, that's when I discover the true horror.

No gingerbread. No music. And from the vacant spot in front of the picture window, the tree isn't even up yet and there are only six days until Christmas.

My dad is the first to greet me coming out from his study.

"There she is. Our Stell-bell is home," he sings like the jolly old soul he is.

"It's good to be home." I sink into his embrace. When we separate, I notice that his hair has more flecks of gray than it used to, and the creases at the corners of his eyes are deeper than they were even months ago.

"Where's the tree? And all the decorations?" I ask.

"Sorry, Stell-bell, we didn't have time to get any decorations up this year. With Sadie's wedding and your mom and I getting older, some things haven't gotten done."

Some things? *Some things?*

Nothing has been done. A surge of disappointment rushes through me.

Decorating for the holidays isn't behind, it's not happening at all this year.

Through the empty picture window in the living room, my eyes drift across the street.

The Jensens' house is lit up tastefully in warm white lights. Their twenty-foot spruce could pass for the Rockefeller Center tree with its large bulbs of multi-color lights. All eight of Santa's reindeer are poised perfectly on their snowy lawn with Rudolph and his red nose leading the group, pulling a person-sized sleigh that people take pictures in at their Christmas Eve party. That sleigh is the infamous site of Jasper's and my heated debate about the merits of a real versus artificial Christmas tree.

I'm team real tree, and always will be.

Secretly, I love seeing that familiar reindeer display with Santa's sleigh.

Their house has always looked picture perfect. It's another way Jasper and I have competed over the years, and it's clear that this year he's winning.

My dad directs me toward the kitchen but before we get there, my sister, Sadie, comes flying toward me.

"Holy shit," she exclaims in way of greeting. "We heard the news!"

"Language." My dad attempts a scolding tone, but it comes off more as a question.

Sadie and I are only a year apart. My mom didn't think she could get pregnant while breastfeeding if she didn't have her period. A few months later, that myth was busted and she was pregnant with Sadie. Growing up, many people thought we were twins because of how close in age we were.

"Dad, I'm twenty-seven. I can say shit." She waves him off. "Anyways, this development requires some expletives."

"What is it?" my dad asks, concern in his voice.

"Stella is dating Jasper!" Sadie squeals again like she can't quite believe it. I get it, though, because me either.

My dad looks confused. "Jasper?"

"Jasper Jensen!" Sadie erupts for the third time. "What other Jasper is there?"

"I thought you and Jasper didn't get along?" my dad muses.

"They don't! They didn't! But now they're dating!" Sadie screams excitedly. "Oh my god, Stellie, you have to tell me everything."

"Let her get settled a minute." My dad directs us into the kitchen where we find my mom.

"Stella!" My mom wipes her hand on a dish towel, coming around the island in her flour-covered apron to wrap me up in her arms. The moment she's got her arms around me, my eyes burn with how good it feels.

"Are you making gingerbread?" I ask, sniffing the air but coming up empty.

"No, I've decided to make Sadie and Tom's wedding cake. I've been working on different frosting recipes. Do you want to try some?" She motions toward the bowls on the island.

The thought of tasting sugary frosting makes my already sensitive stomach ache.

"Maybe later."

Daniel is sitting next to Tom, Sadie's fiancé, at the island scrolling on his phone. He glances up at me and for a split second I feel bad for lying to him, but the alternative, two weeks of awkward flirtatious encounters, is far worse.

"We heard about Jasper." My mom gives me a secretive smile. It's one that says she expected it all along and thought our rivalry was some kind of front for our true feelings.

What she doesn't know is that the rivalry between me and Jasper is perfectly intact.

But I need to make sure Daniel thinks otherwise.

"That's right," I announce loudly. "Jasper and I are dating. He's the man of my dreams."

Sadie laughs because she knows me well and right now, I do sound a bit like a robot.

"Okay, Stellie. Let's get you unpacked."

Sadie pulls me and my suitcase past a forlorn looking Daniel whom Tom throws a comforting arm around and mentions they should go play video games.

When we're tucked into the safety of my bedroom, Sadie starts firing off questions.

"How did it happen with Jasper? Did it start with hate sex? Are you in love?"

"Hate sex?" I ponder, because it never occurred to me that you could hate someone and want them to fuck you.

"Yeah, you know when you're so mad at someone that you bang angry. Tom and I have done it a few times, okay only once, but then we pretended to do it again because it was so hot, but it was really hard to replicate. That anger and tension doesn't create itself."

"I don't know." I shrug. "It just happened."

"That's impossible. There's been so much tension between you two there had to have been fireworks. An explosion of some sort."

There had been an explosion on the plane. My stomach contents.

Now that I'm thinking about it, that was really gross and Jasper handled it like a champ. I'd have to say that if the tables were reversed and he was the one sick, I would not have let him put his head in my lap.

"I was sick and he took care of me, so I decided he wasn't the worst person in the world anymore."

"Aww, that's so sweet."

I can't take her hopeful, lovesick face. It's like she beats the truth out of me without even saying a word.

"I'm lying. He's not my boyfriend. Not really. We're pretending to date each other so Daniel doesn't think there's anything between me and him, and Jasper needs a date to his family's Christmas party."

"I knew something was fishy." Sadie sighs, clearly disappointed with the truth.

"You can't tell anyone. Especially not Tom. He'll tell Daniel and it will all be for nothing."

"Yeah, sorry about Daniel being here. Tom couldn't spend Christmas without him, and I couldn't spend it without Tom. It kind of snowballed."

"It's fine. I'll pretend to be with Jasper and get through the next two weeks."

"So how is that going to work?"

I shrug. "I think we'll hang out a few times. I'll go to his family's party, he'll come to your wedding as my date, then we'll part ways."

"That sounds so simple." She hands me another packing cube from my suitcase. "You are aware this is you and Jasper, right? Nothing has ever been simple with you two."

"Yeah, I know."

Later, after dinner, I get ready for bed, then pause in front of my bedroom window, looking out at Jasper's house.

For a moment, I let myself enjoy the beautiful light display, forgetting that it's the house of my nemesis and now, fake boyfriend.

Then, I yank the curtains shut and crawl into bed.

"It's awful here, Pip. Come rescue me," I whine into the screen of my phone where my best friend, Pippa, is staring back at me.

"This doesn't sound like the magical Cedar Hollow you've told me about."

"Sorry, I'm in a mood. And it's not the town's fault." I shift my eyes upward at the mountains around me and take a deep breath. People from all over the world visit Cedar Hollow and the surrounding area for skiing and other winter recreation and I am lucky to have grown up here. I love New York, the hustle and bustle, but when I come home to Cedar Hollow, a charming and cozy mountain town nestled in the heart of the Rocky Mountains, I immediately feel at peace. But the last twenty-four hours have been anything but peaceful.

"That's understandable, considering you're now dating your arch rival." She arches an eyebrow for emphasis.

I told her everything about my plane ride, getting sick and Jasper witnessing it all, then Daniel showing up at the airport, and my genius plot to make Jasper my boyfriend to avoid Daniel's unwanted advances. It seemed like a good idea at the time, but now that I'm living that reality it's a different story.

I'm currently walking to the tree farm to pick out a tree. With my family tied up with wedding planning, I'm taking it upon myself to decorate for the holidays, starting with the most important item. A real Christmas tree.

Does it make sense to walk to the tree farm, maybe not. But I'm counting on them having a delivery service.

I swing the axe in my hand. The one I grabbed from the

garage, when I was sneaking out of the house so I wouldn't come face to face with Daniel.

"I know what would make me feel better. If you came here."

"Can't. I'm on deadline."

"You're always on a deadline."

"Most authors are. It's kind of how this whole writing books thing goes."

Pippa writes spicy romance novels. She's the one that introduced me to romance audiobooks. They're steamy and fun, and sometimes when I listen to her books, I have to pretend my best friend isn't the author so I don't feel weird about how turned on her descriptive sex scenes make me. Her career has taken off and now she's attending signings and doing appearances all over the world. I'm super proud of her.

"It's Christmas. You should take some time off," I argue.

She sighs. "Maybe next year."

"I'll hold you to that."

"Say hi to Jass for me," she says teasingly.

"Ugh, I will do no such thing. He already knows everything about my adolescent life. There's no way I will let him infiltrate my adult one."

Except, he already has. He's my fake boyfriend now.

I end the video call with Pippa, and continue my walk toward the tree farm.

Suddenly, my phone buzzes and I'm hoping it's a text from Pippa telling me she's booked a flight. It's not. It's the devil himself.

> Hey, Sparky. How are you feeling today?

> Never better

> Glad to hear that I nursed you back to health

>> That's not what happened

> It must have been my unreasonably handsome nose that did it

>> Lose my number

> How would we arrange fake dates?

>> I'm going to change my mind about this whole thing

> I'll stop by your house later

>> Please don't

I drop my phone in my pocket and keep walking.

I didn't imagine I'd be headed to pick out my family's Christmas tree alone but that's okay. I'm used to taking charge and getting things done. So, I'll pick out a tree and have it delivered. Everything will be fine.

Only it's not just having the tree, but sharing the experience of picking it out then taking it home to decorate it. The laughter and the reminiscing about the ornaments while we decorate.

That's going to be the part I miss.

But I'm determined to not miss out on this tradition, even if I have to do it myself, so I trudge on down the sidewalk, taking a right onto Pinecone Way and onward toward the Frosty Fir Tree Farm.

SIX
JASPER

I'M DRIVING to the post office to mail packages for my mom when Liam, my best friend from college and current business partner, calls.

"You make it home okay?" he asks.

"Yeah, flight was easy."

"Still don't know why you didn't take the company jet. I mean flying commercial during the holidays? And not even first class. That's disturbing."

"You know why."

"Yeah, and it's fucking sad to watch my best friend desperately in love with a woman who doesn't see how great he is."

Liam knows how I feel about Stella. He used to give me shit in college for not trying with women. The fact is, I did try to get over Stella. I thought college would be a fresh start, a way to move on and meet someone new. But there wasn't another woman like Stella.

Yes, she's gorgeous, nonetheless it's not only the fact she's stunning and lights up a room, but there's an energy about her that pulls me in while simultaneously driving me crazy. Even as kids, she pushed me to be bolder, sharper, and it's partly because of that drive to compete and push myself that helped me achieve success in my field and start my own company.

All these years later, there still hasn't been anyone who challenges me and meets me head on like she does.

In fact, with the success of my company, it's been even harder to date. A good number of women are enamored with the idea of dating a wealthy man. Whereas my success hasn't changed Stella's opinion about me, and while I want to make amends with her and stop this decades long fight, I appreciate that about her.

In my college years, if I was dating someone before the holidays, I would break things off with them before going home. Liam thought it was because I was cheap and didn't want to buy them a Christmas present, but it was because I didn't want to be dating anyone when I went home. Stella was the force that drew me back home every year, clinging to the hope that things between us would be different.

Spoiler: it never mattered. Stella was determined to keep me at a distance, so I did the only thing I knew and continued to fight with her.

But things are going to be different this year.

"Stella and I are dating," I announce, smugness oozing from my tone.

"Are you bloody serious? Is she an amnesiac? Did she hit her head and forget that she hates you? Worse yet, did you hit her on the head? Was that part of your plan?" Liam sighs, his British accent making it even more dramatic. "No, don't tell me. I don't want to be an accomplice in this scheme. God damn it, Jas, the company is on track to have the best fiscal year yet and now you're going to ruin it all with a lawsuit and felony charges."

"Cool your jets, L. I didn't hit anyone over the head. Stella and I are fake dating."

There's a long pause before another sigh from Liam. "Okay, I know you've been working long hours, mate, but maybe it is time to follow that mental health program we brought on for all the employees. Burn out is a real thing."

I ignore his insinuation that I've lost my mind.

"Stella needed a fake boyfriend for the holidays. A past hookup she wanted to avoid is staying with her family, so she asked me to run interference."

"Now I'm worried about that felony charge again. Tell me the other guy is still in one piece."

Daniel's harmless, but that hasn't prevented me from feeling a surge of jealousy when I've thought of him and Stella together. The fact that she has no interest in him, and is going to great lengths, by fake dating me, to hold him at a distance is keeping him safe.

"He's still in one piece."

"This whole set up is making me nervous. Do I need to come out there and check on you?"

I chuckle. "Is that your way of saying you miss me?"

"It's my way of saying I need you not to lose your head over this woman. There's no reason anyone should lose their head over a woman."

That's Liam. His charming smile and accent make him popular with the ladies in LA, but he's in no rush to settle down.

"If you want small mountain town charm, you know where to find me."

We hang up and I think about what Liam said. I haven't lost my head over Stella, but my heart is a different matter.

❄

I'm on my way back from the post office, passing by the Frosty Fir Tree Farm, when a flash of pink among the trees catches my eye, so I slow down on the road. A woman is aggressively hacking at the base of a tree. I can't imagine that style of chopping is going to get her very far. I can make out the blonde hair trickling out from under the stocking hat. Even from this distance, I know it's Stella.

I pull into the parking lot, then make my way through the rows of trees. For a moment I think I've lost her, but then I spot her pink hat with a large fuzzy pompom on it bobbing around near the base of a giant spruce. The tree must be eight feet tall.

"What are you doing?" I ask.

She pops her head out to look at me.

"Come on, Jensen. It's a Christmas tree farm. Use your brain."

"I know what you're doing. Why are you doing it by yourself?" I glance around, confirming she is alone. "Where are your parents? Where's Sadie and Tom?"

"They're at the final menu tasting for the reception dinner," she says before moving around the tree to find a better angle.

"You didn't want to go?"

"No, I'm busy setting up Christmas."

When I arrived home yesterday evening, I'd noticed there weren't any Christmas lights on the St. James' house. One thing I learned about Stella in the fourth grade is she's obsessed with Christmas light displays. Her family has always gone big on Christmas, but it seems that with Sadie's wedding in a couple weeks, things like outdoor lights and from the looks of it, even putting up a Christmas tree has been put aside.

She taps the earbud fitted in her ear. "Now leave me in peace so I can listen to Taylor Swift's 'Christmas Tree Farm' on repeat while I cut this bad boy down." She turns her back to me and continues chopping at the tree.

"You know, they have people to help with this part."

"What's a matter, city boy? You can't chop down your own tree?"

She's goading me and fuck if she's not damn good at it.

"All right, give me the axe and I'll take a turn."

"You think I'm going to give *you* an axe?" She waves it around, laughing maniacally, which is ironic since she's the one inferring I could be an axe murderer.

"Really, Stella? You think I'd murder you out here?" I motion to the families wandering around. "There are too many witnesses."

"You're so funny, Jasper." She quirks those perfect lips at me. "Funny *looking*."

"That line was played out in middle school. Surely, you've got something better by now."

"I like it just fine. Besides, no one tailors their insults at the request of their nemesis."

"Nemesis? Have I been upgraded from rival?"

A couple passes us, walking hand and hand in search for their tree. They look happy and in love. It could be their first Christmas together. It could be their tenth.

I think about all the holidays I watched Stella with her family across the street. How each time I've returned to our sleepy mountain town, I've been hopeful that things would be different. I realize now how much time I've wasted waiting for Stella to wake up to the fact that we're not kids anymore and I don't want to compete with her. I want to stroll through a field of pine trees to pick out a Christmas tree with her. Then go home and set it up together, using a mix of ornaments from her childhood and mine to fill the tree, alongside some we've collected as a couple. Then, with the glow of the fireplace at our side, I'd drown in her sweet pussy before fucking her so good, she leaves scratch marks down my back. Or something like that. I haven't given it much thought.

That's not reality...yet. Right now, I'm her fake boyfriend, one she despises at that. But fake dating has to be better than being friend-zoned, so I'll take it.

"Come on, Stell, let me give it a shot. How can I earn my fake boyfriend credentials if I don't do something rugged and masculine for you?"

"No. Get your own tree."

She takes a swing but the tree trunk doesn't give at all.

She's holding it wrong, both hands gripping the end like a baseball bat.

"I was looking forward to Christmas with my family. All of my childhood traditions."

Another swing, this one angrier than the next.

Thwack.

"Baking gingerbread houses."

Thwack.

"Driving around to see the light displays. Nobody has time for anything but planning a fucking wedding. Not even a tree." She laughs but it's the sound of defeat. "They put up a tiny fake tree on a table in the foyer." She sniffs. "It's so small it fits on a table, Jasper."

Her eyes lift to meet mine. The cloudy afternoon sky makes for the perfect lighting to see every single one of Stella's freckles. Her skin is fair except the tip of her nose that is pinkened from the cold air.

That's when I realize she's not mad at me, she's upset at the situation. Sadie's wedding is throwing off the holidays for their family and it's upsetting for Stella, the girl who lives for Christmas.

I move toward her, which is quite brave seeing that Stella is wielding an axe and she's hostile. But this time when I reach out a hand for the axe, she relents.

The second it's in my hand I notice something odd. There's not much weight to it and upon further inspection the edge isn't even sharp.

"Stella, where did you get this axe?" I ask.

"I brought it from home. It was in the garage with some of the Halloween supplies."

"Gotcha."

It's a fake axe. A Halloween prop.

In the past, I would have used her gaffe to tease her relentlessly, but even if Stella doesn't know it, this is a pivotal moment for us. Sure, a couple can tease each other about silly mistakes and laugh about it, but Stella's still on the defense with me. She needs to feel safe making mistakes and know I won't go for the jugular and make her feel judged.

It's baby steps with her.

So now I'll need to distract her while I cut the tree down with the saw the tree farm is providing.

"It seems like you need a break. How about I cut down your tree while you go peruse The Cozy Cabin Craft Shop?"

She dusts off her mittens. "I guess I could look around."

"Get anything you want. It's my treat."

She eyes me suspiciously, before turning in the direction of the shop.

"You sure? I have a penchant for holiday gnomes and chunky crocheted scarves I'm never going to wear."

I smile at her warning. "I'm sure. I'm your wealthy boyfriend after all."

She points her finger and gives me a stern look.

"Wealthy *fake* boyfriend."

For now.

"Stella?" I call.

"Yeah?"

"Will you pick me out something, too? Something you think I'd like."

Her face lights up and the wicked gleam of excitement flashes through her eyes.

"Of course, Snowflake," she says a little too sweetly, before turning to make her way to the craft shop.

Seeing the playfulness return to Stella's face is worth every penny of the junk she's going to pick out for me.

I was right. Stella picked out a bunch of random shit for me at the holiday craft shop.

Naughty or nice socks, Santa's lump of coal soap, a "festive farts" candle, and a reindeer butt bottle stopper. Oh, and an ugly holiday sweater. It's green with glittery tinsel adorning the front and cross-stitched Christmas lights sewn across it.

It's nice to know she was thinking about me.

For herself, she picked out one handmade gnome figure. I was surprised she hadn't cleaned out their entire stock, but knowing Stella she probably hated the idea of owing me for anything.

After I got the tree cut and the tree farm worker helped

me bag it, I secured it to the top of my rental SUV—incidental scratch charges be damned—because rideless and determined, Stella had walked to the tree farm thinking they could deliver the tree later. That is not a service the mom-and-pop tree farm offers.

When we pull up to her house, Stella motions to the curb.

"You can pull up here and I'll jump out."

"With an eight-foot tree?" I quip. "I'm going to pull in your driveway and unload it like a normal person would."

"You're always making things more difficult than they have to be."

"Right back at ya, Sparky."

Stella ignores me and starts to untie the tree from the roof. A normal couple could navigate this easily. The tree isn't that heavy, it's just long and awkward to hold. But Stella and I are legendary for our ability to argue about anything. How to get this tree inside her house will be no exception.

After some maneuvering and a few choice words from Stella, the tree along with a ton of loose pine needles is in her family's living room. We work together to get it set up in the tree stand, and arrange it in front of the large picture window.

"It's perfect," she announces, staring at the tree while I keep my eyes fixed on her.

"Yeah."

"Now, I've got to decorate it." She rubs her hands together in anticipation.

"I can help you."

She stares at me a moment, those eyes of hers contemplative, but then she nods her head.

The wool sweater I'm wearing doesn't lend itself to

being pressed up against the pine needles, not to mention the patch of sap that is clinging to my arm after carrying it in.

"I'm going to run home and change my shirt."

"Fine by me." She waves me off.

I rush out the door, anxious to get back before she changes her mind.

SEVEN
STELLA

AFTER RUMMAGING through the attic to find the bins of our family ornaments, I make my way back to the living room, but stop short when I find Daniel walking through the front door. He's in a black zip-up and joggers with a beanie on his head.

"Hey, Stella." He smiles, giving me a flirtatious wink. "What are you up to?"

"Um, hi. I'm about to decorate the tree." With the bin in my hand, I motion to the tree Jasper and I just set up in front of the living room window.

"That's cool." He reaches for the hem of his pullover and yanks it over his head. "I went for a run and now I'm overheating."

I smile. "Yeah, that happens in the Colorado sunshine."

We're making small talk. This is good. I don't want things to be weird between Daniel and me, I just wanted him to back off with any romantic notions about us.

I set the plastic bin down on the coffee table, then head back to the attic for another bin of tree decorations.

"Where did you run? Around the neighborhood?" I ask as I'm walking into the living room, but this time I find Daniel standing in front of me bare-chested and wiping at his neck with his t-shirt. He drops the shirt on the couch and moves toward me.

"I can help you with that."

Before I can respond, he takes the bin out of my hands. It doesn't weigh more than twenty pounds, I had been carrying it easily, yet somehow his biceps manage to flex like he's in the middle of a strenuous set of curls.

He lowers the bin to the ground and I swear he's sticking his butt up more than necessary. Like he wants me to check out his ass. It's a male version of the 'bend and snap.'

My lips press together, holding in a laugh.

I've never seen peacocking before, but I think Daniel's display of muscles and masculine energy is supposed to woo me. It doesn't work. Not because Daniel isn't attractive, because he is. The best way I can explain it is he's like Joey from *Friends*. Sweet, attractive, and funny, but there's not much substance there, at least not between the two of us. He'd make some woman very happy, but that woman isn't me.

He glances between me and the bins filling the living room. "This is a lot of decorations. Do you need help?"

"No, I can manage on my own."

I don't want to decorate the tree by myself, but I'll do that before I decorate it with Daniel. Then, I remember Jasper ran home to change and he'll be back to help.

"You sure?" he asks, giving me a megawatt smile as he stretches his arms upward, making his abdominals contract. "I've got a long wingspan. Good for reaching high things."

"Thanks, Daniel, but I've got it from here."

I pull out a strand of lights and using the ladder Jasper got out for me, I start at the top of the tree.

In my haste to appear busy so that I can ignore Daniel, I don't check that the ladder is locked into place. When I lean to reach around the tree, it pops open the rest of the way and I launch forward. My arms extend in anticipation of protecting my face from hitting the floor, but at the last second, Daniel catches me around the waist.

A rush of relief hits me that I didn't hit the ground, but at that very moment, Jasper walks in the front door.

"I'm back. Do you—" Jasper stops short, taking in the sight of Daniel and me.

I don't know exactly what the visual is, but from what I know, Daniel has his arms wrapped around my waist, and he's shirtless. It might not be interpreted well. Even by a fake boyfriend's standards.

"What the fuck is going on?" Jasper growls.

I don't know if it's the deep growl that emits from Jasper's chest or the fierce look of protectiveness in his eyes, but my body reacts with an intensity I've never known.

My heart hammers against my ribs and my breathing quickens. What is going on?

Daniel, for his part, doesn't panic enough to drop me, but makes sure I'm steady on my feet before he lets me go.

"I was helping Stella."

Jasper narrows his eyes at Daniel. "Without a shirt on?"

"He went for a run and was all sweaty," I offer, but Jasper doesn't relent. He doubles down with the jealous boyfriend act, his jaw clenching as his molars grind together.

For not getting the hint I wasn't interested in him earlier, Daniel is clearly picking up on the situation now.

"I'll let you two finish up here." Daniel grabs his shirt

from the couch and moves toward the stairs. "I'm going to go take a shower."

"Hey, Daniel?" Jasper calls over his shoulder.

"Yeah?" Daniel responds from halfway up the stairs.

"Do me a favor, and don't think about my girlfriend while you're in there."

My eyes bulge in disbelief. No, he did not say that.

"Jasper," I admonish, but I'm not quite sure why. It's kind of sweet that he is so blatantly protective of me. Also, it's making my insides flutter.

"What?" he asks, pulling off his coat and hanging it in the closet.

"I can't believe you said that."

"I can't believe I walked in here and he had his hands all over you."

"He caught me when I fell off the ladder. It might have looked odd, but it was innocent."

Jasper's brows lift. "And he was shirtless?"

I scrunch my nose. "Yeah, that part was weird."

You know what else was odd? How turned on I got at the sight of Jasper jealous. Or at least pretending to be jealous. Since he's my fake boyfriend and it wouldn't have been very convincing to Daniel if Jasper had no issue with him being shirtless with his arms wrapped around me.

The absurdity of the situation hits me and I can't contain my laugh.

"What's so funny?" Jasper asks.

"You. Daniel. The whole situation. I'd say I'm impressed that you managed to play the jealous boyfriend. Maybe your acting skills are better than I thought."

He doesn't say anything else, just stands there looking at me like I exasperate him.

"You don't have to stay and help me decorate. I think Daniel got the point."

"I'm staying."

"Fine. You can finish up the lights, while I get the ornaments ready."

Jasper resets the ladder, and gets to work on hanging the lights on the tree while I open the ornament bin.

Seeing the familiar ornaments peeking up from the soft packing tissue has my face lighting up. All our family ornaments have a name and a year written on the bottom. There are ones from my parents' childhood, then a few from their newlywed years before Sadie's and my collection took over the bulk of them.

Carefully, I take out each one, setting them gently on the coffee table.

There, at the bottom, I pull out a familiar clay snowflake and a rush of nostalgia hits me.

The clay is shaped into six lumpy branches each with a design of dots, lines, and swirls carved into them by toothpick. The original white clay is covered mostly in an icy blue paint with a bit of the white edges remaining, and after twenty years, the silver glitter that once was plentiful is diminishing. A sparkly silver ribbon hangs from the top of the ornament. It's the craftsmanship of a second grader, but even all these years later, I can see the care that went into it.

It's the ornament Jasper gave me the first Christmas after his family moved to Cedar Hollow.

I turn it over. The unfinished back of the snowflake claims what it always has in wobbly penmanship.

For Stella. You're cool like this snowflake. Jasper.

My eyes shift to the man hanging lights on my family's Christmas tree. His thick, coppery-brown hair. The way his jeans hug his ass perfectly.

Jasper turns to find me staring.

"The lights are done. You ready for ornaments?" he asks.

EIGHT
JASPER

WHEN I TURN around to tell Stella the lights are on, she's staring at me with a far off look in her eyes.

"The lights are done. You ready for ornaments?" I ask.

"Um, yeah. Just a sec."

My eyes land on the snowflake ornament in her hand. I recognize it immediately. I made it for Stella the first Christmas after we moved in across the street. The supplies had been offered at a holiday crafting hour my mom took me and Juniper to at the library. While most kids made one for their family, I'd made one for Stella, the girl across the street that I had become enamored with since the start of the school year.

I take the ornament from Stella's hand and flip it over.

I'd thought the inscription was clever. Calling her cool because it was a snowflake and snow is cool. I was a nerdy eight-year-old with a crush.

Seeing it in her hand, knowing she kept it all these years, makes me happy, I can't help but smile.

"I figured this was at the bottom of the landfill."

"Unlike some people, I don't destroy others' art, even if it is mediocre and messy."

She snatches it back.

I know what she's referring to. An incident between us in seventh grade when I bought her drawing at the winter art fair, then told her something that wasn't true. But I want to keep things light between us.

"Is that it, Stell? Or did you keep it because you like me?" I'm teasing her, but I want it to be true so fucking bad.

Her nose scrunches at my suggestion. She's so beautiful it makes my chest ache.

"Because that's why I made it for you."

It's a small truth I can offer her. Telling her I liked her when I was eight is less vulnerable than saying I'm in love with her, and have been, for a long time now.

At my words, her face goes slack. Gone is the pinched look of annoyance, and in its place is something else. Curiosity.

I don't know how it happened, but our bodies have drifted closer to one another, only the smallest space between us. Staring down at her, our eyes are locked, our lips parted slightly in anticipation.

My head dips, closing the space between us, but I pause, refusing to take something I'm not being offered. But then, Stella's hand finds its way up my chest, her fingers sliding in behind my neck, and everything hangs on this one moment.

Our lips are milliseconds from touching. I can feel the warm, minty pant of Stella's breath against my mouth.

Somewhere, a door slams shut.

At the sound of voices in the hallway, Stella pulls back.

I'm still in a daze, having been seconds away from kissing Stella, when her parents appear in the doorway.

Stella's mom's eyes light up when they land on the tree, and then even more when she sees me and Stella.

"Jasper, what a surprise!" Suzanne St. James pulls me in for a hug. "We haven't seen much of you at our house since you and some of your friends toilet papered it senior year."

"That was a mess to clean up." Stella's dad gives me a stern look.

"Yeah," I scratch the back of my neck, "sorry about that."

"It's no matter. The past is in the past." She smiles. "It's nice to see that you and Stella have put all that hostility behind you."

"Yes, we have. Isn't that right, Sparky?" I tease, trying to lighten the mood between us but Stella is completely zoned out.

Stella's mom glances between us as if to say, *okay, children, go ahead and show me.* I wrap my arm around Stella's waist, then press a soft kiss to her hair. She stiffens against me and I know she's doing everything in her power not to elbow me in the ribs. Whatever had passed between us a moment ago was chased away by her family's appearance.

Her mom clasps her hands together. "You two make such a beautiful couple. Maybe there's another wedding in our future."

"Over my dead body," Stella murmurs.

"What was that?" her mom asks.

"Nothing," she says.

"I'm going to fire up the grill," her dad says, turning to head to the back patio.

"Jasper, would you like to join us for dinner?" her mom asks.

"He can't," Stella interjects, her eyes widening to relay the message that I'm not welcome at dinner.

I toss her a teasing smirk. "I'd love to join. Thank you."

Behind her mom's back, Stella narrows her gaze and bares her teeth in my direction. The softness from before is nowhere to be found.

I wonder if she'd really bite me. I'm so fucking crazy about her, I'm sure I'd like it.

"Oh, and the tree looks wonderful." Her mom turns back to address us.

"I've still got to put the ornaments on," Stella says.

"I'll leave you to it." Her mom walks off into the kitchen.

Stella turns to me. "You don't need to stay for dinner. Your help with the tree is convincing enough."

"I better stick around. Daniel will be freshly showered and might get some wild ideas."

"Hmm, maybe. Do you think it's weird he's still flirting with me even though he thinks I have a boyfriend?"

I take the ornament from her hand and hang it high up on the tree.

It's never stopped me.

"That's why it's important we show him our relationship is unshakable."

"Yeah, I guess you're right."

A huge smile pulls at my lips, and Stella scowls at me. "Don't let it go to your head."

"I'll try not to."

Stella turns on her holiday playlist and we decorate to the tunes of Mariah Carey, Kelly Clarkson, and Britney Spears, our almost kiss gone, but not forgotten.

NINE
STELLA

"EVERYONE IS GOING to The Merry Moose tonight to hang out," Sadie announces as she passes the platter of beef tenderloin to Daniel.

"Who's everyone?" I ask, distracted as I pull my eyes from where I've been staring at Jasper's lips for the hundredth time this meal.

I've been replaying the moment between us by the tree over and over, but it still doesn't feel like it could have happened. Did I almost kiss Jasper?

His head had dipped low, making his intentions known, and instead of pulling back, I leaned in.

That wasn't even the most alarming thing.

In that moment, when I was teasing the back of his neck with my fingers, I realized I'm attracted to Jasper. There had been the tell-tale flutter in my belly, the quickening of my pulse, and then to my horror, my underwear was soaked.

After my parents interrupted us, I'd pretended nothing had happened, continuing to hang ornaments with Jasper on the tree, all the while fighting the dull ache between my thighs. Before dinner, I'd escaped to my room to change my underwear, vehemently tossing the wet pair into my laundry hamper to rid myself of the evidence of my body's traitorous actions.

"Friends from high school, anyone in town for the holidays and the wedding." Sadie's response brings me back to the conversation.

That sounds like a bad idea. It's one thing to try to convince Daniel, who doesn't know Jasper's and my history, but I don't know that I'm ready to take our fake relationship out in such a public place. Things between us are intense, maybe now more than ever, as I try to figure out what that almost-kiss means. How are we going to convince our old classmates that we've patched things up and started dating of all things?

Before I can turn down the invitation, Jasper answers for me.

"We'll be there."

The Merry Moose is a favorite hangout spot in Cedar Hollow. Its décor changes with the season and my favorite time of year is when it's decked out in holly and strands of lights with large, vintage-style colored bulbs. Wreaths are hung on the walls, ornaments dangle in rows from the ceiling, and the large mahogany bar is lit up red and green beneath the bar top.

In the back of the bar there are two pool tables, a dart board, and a Skee-Ball game that I currently hold the record for the highest score on.

The moment I walk through the front door, it's cozy and warm, with Christmas music playing beneath the buzz of people's conversations. I'm happy I came.

Jasper walks in behind me, and helps me out of my coat, hanging it on one of the many hooks on the wall by the door.

After I adjust my sweater and hair, I turn back to the bar and the whole place is staring at us.

We're like exotic animals at the zoo. Jasper and Stella out in the wild. Never seen before. Or at least not like this.

Jasper clasps my hand in his and pulls me forward. "Come on, let's get a drink."

Our forward motion snaps everyone back into their conversations yet I can still feel eyes on us as we make our way to the bar.

Beside me, Jasper's shoulder brushes mine. The contact shouldn't be unnerving. It's a simple shoulder brush. Two sweaters rubbing up against each other. But combined with the depth of his hazel gaze on me and the way his scent—warm, yet fresh like mint and cool mountain air—wraps itself around me, I'm hyperaware of how every cell in my body is reacting to him.

I fight against every single one of those cells, and stiffen in response.

"Why are you so close?" I ask, annoyance seeming to be my only way to handle this new development.

"When you like someone, you want to stand close to them," Jasper whispers.

I cough out a laugh. "So, you can see why this would be a struggle for me."

His confident smile doesn't waver. "Then we'll practice."

Jasper shifts his body behind mine, caging me in against the bar.

Maybe if I suck in, I can create more space between us. But suck in what? My front is already pressed firmly against the bar. My butt?

I squeeze my cheeks together, but the action only makes my glute muscles swell, lifting and extending my ass outward. The opposite of my desired outcome.

Sucking in your butt isn't a thing because it's just there existing behind you, and the only thing it can do is flex or relax.

"Are you flexing your butt?" he asks.

"You wish," is my only comeback.

What the hell is wrong with me? I'm a twenty-eight-year-old woman who has stood next to men before.

Not Jasper Jensen, my oh so helpful brain retorts.

So, I give up and let my butt exist in the miniscule space between us.

He lifts one hand off the bar to get the bartender's attention, and the shift in movement has his front brushing against my back, his black denim-clad crotch skimming against my ass. It's only a graze, and yet my body reacts like he put his hand between my thighs.

Jasper Jensen, my longtime rival and the bane of my existence, has stirred something inside of me, and that something is now on the outside.

I'm hot. I'm achy. And my underwear needs to be changed. AGAIN.

I'm so startled by the sensation, I drop the glass of water I'd just filled from the water station in my hand and it spills all over the bar.

"Shit. I'm sorry." I reach out for a stack of napkins, trying to mop up the water.

Jasper reaches over me to help, but it doesn't help, because it only reminds my body of how delicious it feels to have his pressed to mine.

"I got it," I snap.

"Just trying to help."

"Well, you're not," I hiss, trying to put extra venom into my response. It's my only defense mechanism at this point. "It's my mess and I can clean it up myself." I'm clearly referring to the spilt water, not what's happening between my thighs.

Just then, Cady Cosgrove, a classmate of ours, approaches us.

"I can't believe it." She giggles. "Stella St. James and Jasper Jensen. Together. And you're not being forced like that community service mandate after the chemistry lab fiasco."

"Ha ha. Yeah, that was hilarious." I laugh extra loud, hoping it will cover up all the uncomfortable feelings I'm having.

Jasper and I had the same top grade in chemistry class, and we were determined to outdo the other by finishing the extra credit lab assignment Mrs. Vlasky gave us, but we got in a fight and nearly burned down the chemistry lab. It earned us after-school community service, and we had to clean the lab while our classmate, Leena Basu, ended up completing the extra credit and getting the top grade.

A small group of high school classmates and old friends join our group and we move to a high-top table near the bar, so we can set our drinks on it.

"We've seen all the articles and media about your company, Jensen," Scott Symes says, "but tell us what you've been up to since we saw you last."

I might have been tempted to sneak off, but the second Jasper starts talking about his company and the virtual reality program they've built for education purposes in schools, I'm captivated.

The program sounds great, but it's also how passionate Jasper is when he talks about it. Several people ask him questions about the technology and that's when he really geeks out. But instead of being turned off, I find myself leaning in to hear more.

The way his eyes light as he talks, the curve of his mouth is not only sweet, but kind of sexy. And before long, I'm staring at his lips again.

The group quiets and I lift my eyes to Jasper's to find him staring back at me, a playful smile on his lips.

"I could go on for hours, but the real excitement is Stella's promotion to Creative Director at East & Ivy." He pulls me in close, pressing his lips to my temple and from the stirring in my core, it's clear I'm going to have to do laundry soon or I'm going to run out of clean underwear.

I give Jasper a small smile. I need to play the part after all. "Thanks, Jass."

"That's so amazing." Cady beams at me. "I love East & Ivy."

"Thank you." I carefully lift my espresso martini in a cheers motion.

"You two are Cedar Hollow's ultimate power couple." Rex, an old cross-country teammate of ours, tips his head toward Jasper. "Congrats, man."

"How do you two do the long-distance thing?" That's Daniel, who just joined the group with Tom and Sadie.

Of course he's the dissenter here. We'll just have to make some story up about video chats and planned weekends to visit each other.

"I'm moving to New York next month," Jasper announces right as I take a sip of my espresso martini and I end up spraying it out of my mouth.

"Jesus, Stellie." Sadie wipes at her face. "What the heck?"

"Sorry, it went down the wrong pipe."

"No, it sprayed out of your mouth."

"Anyways, Jasper is moving to New York?" Cady's eyes light up. "For Stella? That's so romantic."

"For business," Jasper corrects. "And for Stella."

The group continues their conversation, the discussion moving to Sadie and Tom's wedding.

"Excuse me a moment." I pick up my martini and move away from Jasper.

Sadie's in mid-conversation but her eyes meet mine. *You okay?*

I give her a quick nod before leaving the group. I can't talk right now, I need a moment to collect my thoughts, so I head straight for the back of the bar where the Skee-Ball game is located. A game of Skee-Ball always puts things into perspective.

I pull out my credit card and swipe it on the card reader, then pull the lever to release the set of balls.

Before the game begins, the lights on the scoreboard flash with the initials of the top three highest scorers.

I blink. The top place holder isn't SS for Stella St. James like it was a year ago. Staring back at me in the top spot are the initials JJ.

Jasper Jensen.

TEN
JASPER

TONIGHT, Stella is dressed in a fuzzy black sweater that looks like a cloud, and jeans that hug her ass perfectly. Her hair is styled in soft waves that make me want to reach out and run my fingers through it. She's wearing large jeweled Christmas tree earrings that shimmer in the bar lights every time she turns her head. They're similar in style to the wreaths she had been wearing on the plane.

She's fucking gorgeous, even when she's fuming mad, and right now, she's furious.

I'd given her a moment to retreat after the New York news had settled over the group. Everyone thought it was sweet that I was making our long-distance relationship work by moving to New York. Everyone except Stella.

Now, I've made my way to the back of the bar where she's standing at the Skee-Ball game, her face flush with anger.

She points at the Skee-Ball machine where the top score listed has initials JJ. That's me.

"I can't believe you took my Skee-Ball record." Her mouth gapes open in disbelief.

"I didn't take it. I beat you fair and square." Over the years, I've gotten good at the game to stay competitive with Stella. "You're a talented Skee-Ball player, I'm sure you'll be back on top in no time."

Her eyes light with her signature feisty spark and I know even before she opens her mouth that she's going to make me pay for it.

"You're right, I do have many talents, Jass." She licks her lips, then nods toward the table across the bar. "Just ask Tanner Windell."

I follow her gaze to find Tanner and a few of his friends seated by the pool tables. While I hate that Tanner has touched Stella, I know he's not current competition.

"Tanner Windell is an idiot."

"Doesn't mean he wasn't a good kisser."

"What was that, eighth grade? Jesus, Stella. I hope you've been kissed properly since then."

But I don't mean a word of it because no guy wants the woman they've been in love with half their life to be kissed by another guy. I'm still annoyed by the position I found her and Daniel in earlier this evening. She said it was innocent, an accident, and I believed her, but that didn't stop my blood from heating with jealousy.

"Of course, I have. Lots of times. Lots of proper kissing." She abandons the Skee-Ball game, marching off with her espresso martini to a nearby table, and I follow. "Just so you know, dating in New York City is harder than you think it would be. There are millions of people. Finding the right person is like looking for a needle in a haystack."

"Maybe you're looking in the wrong place," I challenge.

"Maybe you're judgmental and condescending."

I move to redirect the conversation.

"Thanks for the tip about dating in Manhattan. I'll keep that in mind when I move there next month."

"About that. Did you tell everyone that to mess with me? I thought you were there for business. A meeting or something?"

"My company is opening an office there." I watch Stella's face pinch like she's tasted something sour. "Don't look so disgusted. There are millions of people, right? I'm sure we'll never run into each other."

It's not what I want, but it's clearly what she thinks should happen. The clash of how I feel about Stella and the charade I have to keep up to fight with her is starting to exhaust me.

"You're single, too," she retorts.

"You're right. While I've been busy building my company, my personal life has taken a backseat. I plan to rectify that soon."

"I'm sure she'll be lucky to have you." She bats her lashes mockingly. "And I've been busy, too."

"I know. You're the highest paid creative director in the lifestyle brand industry."

"So, you've been keeping tabs on me?"

I give her a cocky smile. "I keep my ear to the ground."

She takes a sip of her drink, assessing me.

"Stella, your career is impressive."

"But?" she prompts.

"There's no but."

"There's always a but with you."

I know I'm being baited. It's what we do to each other. Even when it's honest feedback, it's impossible to see the other person as a sounding board. Our egos are too involved. At least they have been in the past, and Stella's still is.

"I mean it. You're amazing at what you do. I have no notes."

"As if I'd take advice from you. Maybe if you tried to sound like less of know-it-all when you talk to people, you'd have better luck getting a date."

We're treading into hot water and I hate it. I wanted tonight to be different. I'd hoped it would be different, but it's clear Stella doesn't want it to be. She likes the old us. Slinging barbs until one of us runs off licking our wounds. Fake dating isn't changing anything.

"That's not how I meant it. I think you're an amazing artist and creative. East & Ivy is lucky to have you. I'd always envisioned you doing your own thing. Starting your own design firm or doing freelance collaborations."

At my words, her mask slips for a moment. It's like I've hit a soft spot in an otherwise hard shell. But it's only a fraction of a second before her shields go up again.

"I don't need you envisioning me doing anything. In fact, keep me out of your thoughts, okay?"

"That's impossible."

"Well, try."

We're at an impasse and I know I shouldn't push her anymore tonight.

"I'm going to use the restroom," I tell her, abandoning my empty beer at the table.

"Fine." She takes a sip of her espresso martini and reaches for her phone.

After I use the restroom, I take a moment to cool off. This is not going the way I wanted it to. One minute Stella's looking at me like she might actually see me, like the way her eyes softened and her mouth spread into a sincere smile when I was talking about my company and the technology I designed, then the next she's angry about me being one of millions in the city where she lives.

When I come back from the restroom, Stella isn't at the table anymore. My eyes scan the bar and find her talking to Tanner and some other locals.

"You ready to leave?" I ask upon my approach, trying to keep my cool.

Stella takes a step away from Tanner's table, but she barely looks at me before responding. "Sadie and Tom are still here. I think I'll catch a ride with them."

"What about Daniel?" I ask, searching the crowd.

"He's talking to Cady Cosgrove and he looks smitten." She smiles, excitedly bouncing on her toes, "I think I might be in the clear." She grips my forearm, her eyes alight with joy. "If Daniel finds someone else, we might be able to break up, or at least not hang out as much. I mean I wouldn't tell him we're broken up, just in case."

My stomach drops at her suggestion. I thought we were making progress this afternoon and now she's looking for an excuse to back out of our arrangement?

"Stella, I told my family we were together and they're expecting you at the Christmas Eve party."

"Okay, fine." She sighs. "I'll still go to that."

"I'd like to take you home."

"We're fake dating but we don't need to spend every moment together. And I want to stay and hang out with Sadie."

"Listen, I'm sorry about Skee-Ball."

"It's fine. I already beat your score while you were in the restroom." She motions toward the game.

I laugh at that, because of course Stella has found her way back on top.

"What I said about your job. If you're happy, that's all that matters."

"And how would you know if I am? All we do is fight."

"I don't want to fight with you. I try to stay calm, but then things get heated and you drag me back in."

"Well, sorry I'm such a bad influence."

"That's not what I mean." I groan at my words being taken out of context again.

From behind us, Tanner pipes up.

"If you two would stop arguing for a second, you'd see you're under the mistletoe."

My eyes lift to the sprigs of green hanging from a single hook in the middle of the door frame. I watch as Stella's eyes find the mistletoe above our heads, before falling back to mine.

"Kiss! Kiss!" Tanner chants, and soon others around us join in.

Kiss. Kiss. Kiss.

They're probably looking for proof that this relationship is real. Everyone here tonight was a mix of curious and shocked that Stella and I are together.

If Tanner wasn't doing me a favor with his obnoxious chanting, I'd be inclined to tell him to shut the fuck up.

Stella and I didn't discuss this part of our fake relationship. Clearly, we'd never anticipated being caught under the mistletoe together, which in hindsight maybe we should have. It's Christmas after all.

"Are you going to kiss me, Jasper?" Stella's gaze falls to my lips then quickly flicks back to my eyes, her full lips twitching with amusement. "Or are you too afraid?"

Her teasing words ignite the fire in me.

Does she have any idea how long I've waited for this moment?

Every muscle in my body is coiled tightly, but no, I'm not afraid. On the other side of our first kiss, I'm already planning a million more. She's the one who should be nervous.

"Why would I be afraid, Stella?" I ask.

She shrugs. "Maybe you're out of practice. How long has it been anyways?"

"Doesn't matter. All that matters right now is that you want me to kiss you. Do you want me to?"

"Oh, I want it." She smiles smugly. "I want to know exactly how bad of a kisser you are so I can tease you mercilessly."

"Okay."

My right hand reaches out toward her hip, my index finger hooking into one of the belt loops of her jeans and with one firm tug, she's flush against me.

The movement, or the quickness of it surprises her and she lets out a soft gasp, her hands lifting to my chest to brace against the impact. Beneath my palms, Stella's body is soft and supple. I'm dying to trace the line of her curves with my hands but I need to focus on the task at hand.

Kissing her plush mouth and tasting her.

The kiss that eluded us earlier is going to happen now.

Lowering my head, my left hand cups her jaw.

"Last chance, Stella."

"For what?" she whispers, searching my face, as the concern in her expression grows.

"To not know what it's like for my mouth to claim yours."

"Jasper."

My name on her lips is a catalyst. I can't hold back any longer. Tightening my grip on her face, I hold her where I want her and then claim what is mine. Our lips crash together with the force of two opposing magnet ends that have been flipped and finally found their connection.

Our mouths have only been used for teasing and bickering. It's a new dance for us, but I feel the moment Stella gives in. The way her mouth opens to me, and her tongue strokes against mine, inviting me in.

The way she mews into my mouth is obscene. I wish we weren't in the middle of a crowded bar. I wish there was a solid wall nearby that I could press her up against and explore her further.

"Fuck," I growl.

Too soon, the cheering around us trickles in and I know I should pull back before our display becomes indecent.

When I do, Stella's lips are still pressed outward, searching for contact, her eyes still closed.

"Stella?"

Her lashes flutter twice before her eyelids lift and she looks at me. "What?"

I comb a piece of hair over her shoulder.

"It may not seem like it, but I've always wanted to see you win. Whether it's Skee-Ball or a promotion in your career."

She's quiet, but gives a slight nod of her head. Her soft, unfocused eyes telling me she's still reeling from our kiss.

I drop my hand from her waist and turn to leave, but then I remember one more thing I wanted to tell her.

"Oh, and Stella?" I brush a thumb across her lips. They're still wet from our hungry kiss. "I don't want to hear about you and other guys. I don't care if it was ten years ago or ten days ago, for the next two weeks, you, and these lips, are mine."

I press a firm kiss against her mouth. It's clear from the

way her lips part and her body leans into me that she wants more. It kills me to do it but I pull away, leaving her there, breathless and aching for more. It's how I've been living my life for years, Stella can feel what it's like for one night.

ELEVEN
STELLA

JASPER JENSEN IS A LOUSY KISSER.

Sadie takes the permanent marker out of my hand. "Feel better?"

I open the stall door and move toward the sink to wash my hands. My duty is now done in all three stalls of the women's restroom at The Merry Moose. It's a public service announcement. All women should be warned.

Jasper Jensen is a lousy kisser, so no one should kiss him...but me.

"No."

Jasper had warned me. He'd asked if I wanted him to kiss me and I'd naively agreed. I'd been certain it would be a meaningless kiss. Another one to toss on the pile of wasted kisses. Another frog that wouldn't turn into a prince.

But just like his science fair project our sophomore year when he built a wind-powered cell phone charger out of a clock radio, he'd delivered. I'm certain he's at home basking in the glory of having implanted an unforgettable kiss that will torture me until the end of my days.

And then he'd told me how amazing he thinks I am. Can you believe this guy? What the fuck is his problem?

"All right, Stellie. Let's get out of here." Sadie links her arm in mine then guides me out to her car. After Jasper's and my explosive kiss, Tom took it upon himself to be Daniel's wingman tonight. I really hope Daniel finds someone new to project his feelings onto. It would take the pressure off Jasper's and my fake relationship. Then Jasper and I wouldn't need to be spending so much time together. And he wouldn't be required to kiss me under the mistletoe.

Sadie guides me to her car and tucks me into the passenger seat.

I really should put my phone away, but I'm unhinged and horny and just tipsy enough to send Jasper a text.

> That kiss wasn't remarkable but I'm willing to give you a second chance.

He replies right away.

> Our rule has always been no do-overs.

> I made up that rule, so I'm allowed to change it.

> I don't know, Stella. Changing the rules feels dangerous.

Now I'm annoyed. He used my full name in a text situation. That's an extra word he had to type out. Not to mention it stounds like he's scolding me. Also, I hate that I offered to change things between us and he pretty much threw it in my face.

It's a short drive home from The Merry Moose, but I

stew over Jasper's text the whole way. Once we're home, I slam the car door and Sadie gives me a disapproving look.

"Chill. It's a car."

"Not your car."

"Sorry. I'm just annoyed with Jasper. It's like he shouldn't kiss me like that and then not offer to do it again. It's rude and it's just like him to be playing with me."

"Yeah, your hot, wealthy boyfriend who offered to take you home is a real jerk."

"Hey, I wanted to stay and hang out with you. Besides, there's no way I could have left with him after that kiss. It would have meant he won."

"Won what, Stellie?" she asks.

I open my mouth to answer her question, but I pause when I come up empty. It's not a debate or a cross-country race, there's no class honors or student council position to vie for, which makes this situation between Jasper and me feel unsettling. Thirteen-year-old me would have had all the answers, but I don't know how to navigate the adult version of us.

"I'm not sure," I half mumble, "but it's always something with Jasper."

Sadie's face breaks into a Cheshire cat smile.

"What?" I ask.

She clears her throat and nods behind me.

I turn to find Jasper standing at the end of our driveway still dressed in the black jeans and holiday sweater I picked out for him earlier. That stupid fucking sweater. I'm going to make him wear one every day to fend off all the other women he could possibly want to kiss. I'm greedy like that. Don't tell Santa.

Behind me, I hear the front door click shut and glance back to see Sadie has gone inside.

"You didn't have to lie. You could say the kiss was phenomenal and you wanted to do it again."

"I didn't lie," I insist, but we both know my pants are on fire.

"You said it wasn't remarkable." His lips twitch, a confident smirk playing at the corners of his mouth.

"I mean what is remarkable anyway?" I tease. "Extraordinary. Amazing. Notable." I throw up my hands. "They're just words."

He turns to leave, hands tucked into his pockets, obnoxiously nonchalant.

He's going to make me say it.

"Jasper, wait." I groan like a petulant child because that's what he brings out in me. "I need to know."

He stops walking and turns around. A moment later, he's right in front of me, so close I can feel the warmth from his body and it makes me shiver. Or maybe that's because flecks of snow have started to fall from the sky and have landed on my face.

"What do you need to know, Stella?" Jasper asks.

For once, I swallow my pride and tell him the truth.

"I need to know if it was a fluke. A good kiss in the moment that can't be replicated. It's data collection. For science."

Okay, maybe it's only part of the truth.

Fine. One-quarter.

But outright telling Jasper that I want his lips on mine for no good reason other than it felt fucking fantastic is not going to happen.

"What are you going to do with the information?" he asks. "This data collection."

"I don't know yet. I'll need at least five to ten business days to process."

"We don't have that kind of time, Stella." His brows pull down and his lips pull into a firm line. He looks so serious, like this is a project he could get on board with if not for the time constraints.

"Fine," I relent. "I don't know what I'll do with the information, but I need a bigger sample size." I'm getting whiny again, which I hate, because it makes me seem desperate. I am desperate, but again, that's something Jasper can never know.

He steps closer, one hand slipping through the front of my open coat and wrapping around my back. It feels so good to have his hands on me again.

A moment later, with Jasper guiding me, we're moving backwards until my back is pressed flush against the brick of the garage wall.

"You want another kiss?" He asks.

I swallow and nod.

Jasper's face inches closer, but I keep my eyes on him. I need to know what kind of trickery he's got up his sleeve.

"Close your eyes, Stell."

I don't plan to, but the impact of his lips on mine a moment later has my eyelids falling closed.

This kiss is different than the one in the bar, and my brain logs every distinction.

Soft, yet all consuming. Like fire melting new fallen snow. The heat of a roaring fireplace with the gentleness of the fleece blanket you'd curl up next to it in. And all the wicked delight of sneaking the holiday chocolates your mom asked you not to touch.

I wrap my arms around Jasper's neck, pulling him deeper into me. Desperate to dive beneath the surface of this sweet kiss.

Like the strike of a match, the softness is replaced with urgency.

Our hands claw at each other's clothes, our pelvises grind against one another. Jasper's cock is rigid and greedily seeking friction against my stomach.

Now that's a new development.

Jasper pulls back, his hands still firmly holding my hips but where they had been inviting me toward him a moment ago, now they are pressing me backward against the brick.

We're both breathing so heavily, among the soft flakes of snow falling, puffs of air from our exhales swirl between us.

"Holy shit. What does this mean?" I ask.

Data collection is complete and all signs point to me being absolutely, one-hundred percent no denying it, attracted to Jasper.

"You need to sleep on it. But you want to know my theory?" he asks.

"Hmm, I guess."

"I think wrapped up in all that hostility and loathing you have for me, there's a part of you that likes me."

"Whoa." I hold up a hand. "Don't flatter yourself, Jensen. I think I'm just horny."

My body shivers, and he rubs his hands up my arms in a warming gesture. But I'm not cold. I'm so fucking warm, I think I might burst into flames.

If I thought the kiss at the bar was going to keep me up all night, now there's been rubbing and friction. I'll never sleep again.

"I've got a thing to go to tomorrow. Come with me."

"What kind of thing?" I ask, distracted, my mind wandering to the fact that we're both staying at our parents' houses with various other guests and I'm wondering how I'll manage to make out with my fake boyfriend when there's zero privacy.

"You'll like it. I promise."

I promise.

I don't think Jasper has ever said those words to me before.

Tonight is a night of firsts.

Kisses and promises.

It's scary to be seeing this side of Jasper, but if I admit that then I wouldn't be any good at my ability to adapt to this new game we're playing.

"Okay."

He presses one last soft kiss to my lips before he leaves, but I don't go inside. I watch him walk across the street and onto his porch.

"Go inside, Stella," he calls.

I cup my hands around my mouth to project my voice. "You first."

"We'll do it at the same time."

"Fine."

We open our doors and step inside, slowly closing them to make sure the other person is following through with it. When the door is latched, I take a moment to lean against it and collect my thoughts.

Tonight proved one thing, it's a new competition for Jasper and me. With new tactics and rules. But the objective is the same, to stay vigilant, and never drop my guard.

TWELVE
JASPER

I CLOSE the door behind me and slip off my shoes. I'm about to head upstairs when I hear a sound in the kitchen.

I find Juniper there making a snack.

"PB&J with potato chips. You want one?" she offers.

"Hell, yeah." I pull up a stool and start making a sandwich. Once the peanut butter and jelly are spread, I put a handful of kettle cooked potato chips on one side, then layer it with the other piece of bread.

"What's really going on with you and Stella?" she asks. "I thought this wasn't a real relationship, but I just saw you kissing her outside."

"It's complicated."

She laughs. "Hasn't it always been with you two?"

"Unfortunately, yeah. What's been going on with you?" I ask, feeling like we haven't had much time to catch up since I got home.

"I had finals last week. The next month is finalizing my business plan for the bookstore before the spring semester starts."

Juniper reads a ton of books. Like one every day. It's

fascinating. She's got a huge following on social media and posts about romance books there. Now she's months away from opening her own romance bookstore here in Cedar Hollow.

"That's awesome."

"The problem is I'm great at business classes, but applying it in the real world feels daunting. Do you have any advice, from one entrepreneur to another?"

"Surround yourself with people who know what they're doing."

She laughs. "There's no budget for that, yet."

An idea forms in my head.

"You should talk to Liam. He knows all about the business side of things. I'd be lost without him, as a business partner and a friend."

"It's fine. I'm sure I'll figure it out."

"If you need anything, you'll let me know? Financial backing. Tech support. Anything."

She nods, but we've already had this conversation and she wants to do it all on her own. She's ambitious, but I know exactly what that's like. I was doing the same thing when I was her age.

We finish our sandwiches and make our way upstairs. I give her a hug at the landing, then we part ways to go to our rooms.

"Hey, Jas," she calls, stopping at her bedroom door. "I have confidence you and Stella are going to work out."

"How do you know?" I ask.

"Because you deserve an HEA."

"HEA?"

"Happily Ever After."

"Thanks, Juni. I sure hope so."

"I think something about your kiss altered my brain chemistry. I've never had a sex dream before last night."

I slam on the brakes.

"Jesus, Stella. That's not what I was expecting at seven-thirty in the morning."

She shrugs before taking a drink of her coffee.

"We've always been brutally honest with each other, so I figured there's no point stopping now."

"That's true."

"Thanks again for the coffee. How did you know what I wanted?"

I give a casual shrug, wondering if I'm prepared to be as honest as Stella. Her boldness comes from a place of surprise, but the feelings I have for her have been buried under a rock for over a decade. They're not as easily coaxed into the light when I'm uncertain if she's ready to hear about them.

"It's a pretty generic order."

"A Venti Caramel Macchiato with two pumps of vanilla, extra caramel drizzle, three shots of espresso, almond milk, a pump of toffee nut syrup, topped with whipped cream and cinnamon powder?"

"Okay, I asked Sadie last night. I wanted it to be a surprise."

"I'm surprised you went to that much effort. I'd expected you to get me a black coffee and tell me it's the color of my soul."

I lift my cup in her direction. "Touché."

She laughs, loud and carefree. And fuck if I'm not enraptured by the moment.

Stella St. James laughing in my passenger seat is something I could get used to. The last time she was sitting there we were negotiating a fake dating arrangement.

She's dressed in dark jeans and a cropped red sweater. Her blonde hair is tied halfway back with a big white bow. Another pair of glittery holiday earrings hang from her lobes. Today they're candy canes coordinated with her sweater and bow. She looks like a candy cane. She looks good enough to eat.

I'm debating asking her details about her sex dream because we might end up in a ditch. That would be unfavorable with the day I have planned. All I know is it better have been me servicing dream Stella or I'll fucking pull this car over right now and make damn well sure she knows who she belongs to.

"You were saying about your dream?" I coax, knowing I won't be able to focus if I'm running every possible scenario in my head.

"You were in it."

"Damn straight," I mutter.

"What was that?" she asks, distracted.

"Go on."

"You tied me up," her eyes flick over to me before adding, "You were rough with me, but I liked it. A lot. I've never had hate sex but I imagine that's what it was between us. Pent up frustration from all the years of competition."

I clear my throat, the crotch of my jeans much tighter than it was a moment ago. I'm going to need to calm down before we get to the warehouse.

Stella's quiet until we come to a stop at a street light.

"Jasper, have you had sex dreams before?" she asks.

I can't help but chuckle. "Yeah, quite a few."

I pull into the parking lot of our destination.

"Hmm. Any of particular interest?" She wiggles her brows.

I let her question hang between us as I navigate my vehicle to a parking spot and turn off the engine. With an arm draped over the steering wheel, I turn in my seat to look at her.

"Stella, are you asking if I've had a sex dream about you?"

She licks her lips, her boldness from earlier lessening a notch. "Maybe."

"Let me tell you a little secret," I motion her closer with a crook of my finger and she leans in conspiratorially. "They've all been about you."

THIRTEEN
STELLA

I SHOULD HAVE KNOWN that Jasper's idea of a good time would be to get me all hot and bothered, then take me somewhere that it is absolutely not appropriate to be hot and bothered.

Like the Toys for Tiny Hearts fulfillment center where Santa's elves are busy wrapping gifts and spreading holiday cheer to underserved communities and children. I've been given a number and a group to stand with but I'm still not certain what we're doing here.

They've all been about you.

Jasper had delivered that line then promptly opened my door like a gentleman and escorted me inside this lovely holiday workshop filled with the most wholesome volunteers.

I've been assigned to Sandy's group. She's cheerful and sweet and in her sixties. Her innocent demeanor only makes me feel like more of a pervert.

They've all been about you.

What does that mean?

He's had one dream since he saw me this week and it was about me?

How many is *all*? I'm going to need a specific number.

Now, he's across the room, assigned to another group and I'm finding it impossible to focus on what Sandy is saying.

"The rolls are color coded and the ribbons are as well. We've got an instructional video to help you streamline the process of wrapping the presents."

"We're wrapping presents?!" I shriek, throwing my arms overhead in excitement. I'm like Buddy the Elf learning that Santa is coming to visit.

The people in my group turn to stare but I can't be bothered.

I fucking love wrapping presents. Every year I volunteered for gift wrapping shifts at Santa's Workshop at the outlet mall and by my senior year I was awarded the highest gift-wrapping honor...C.E.W. The Chief Executive Wrapper. Santa presented the award and everything.

Jasper knows I was a gift wrapper because the service was free and he delighted in bringing me odd-shaped gifts to wrap. It was just another way for him to provoke me, but little did he know I relished his arrival with irregular-shaped gifts and oversized packages. They were a challenge I welcomed. I couldn't let him know that because it would spoil all the fun, so I pretended to be annoyed and he continued to show up every year with them.

From across the room, I feel Jasper's gaze on me. He's staring at me with the oddest look. Likely because he heard my scream of delight.

Then he smiles and winks. And my panties are immediately wet. I'm starting to notice a pattern.

Which would be irritating if I wasn't immediately

distracted by the fact that Sandy is leading me to my very own wrapping station. The wrapping paper is decadent, the bows are plentiful, and Kelly Clarkson's "Underneath the Tree" starts belting from one of the overhead speakers. This is my dream holiday scenario.

For the next four hours, I wrap presents like it's my job. Because today, it is.

It's not only the joy of wrapping for me, but also knowing it's for a good cause because everyone deserves a beautifully-wrapped present for the holidays.

There's a guide for bow and wrapping paper coordination, but I make up my own and Sandy even compliments me on my combinations.

I show all the others a new technique that helps save on the amount of wrapping paper used but still makes for a pretty wrapped gift.

It's truly the best way to spend a day.

As the morning passes, Jasper stops by my table a few times, supplying me with flirtatious lines like...*can I borrow your scissors* and *do you need me to fill your tape dispenser?*

It's cute. It's sweet. I like this new game we're playing.

On our lunch break, I'm starting to get antsy for some kind of affection from Jasper. He is my fake boyfriend after all so a quick ass grab or boob squeeze wouldn't be out of line.

Or maybe he thinks it would.

We never discussed touching or kissing or anything I want to do with him now, because when we made this arrangement, we were enemies. Technically we still are, but the kiss last night made me aware that I'm insanely attracted to him. Maybe the kiss was some silent truce of our minds so that our bodies could have some temporary fun. Because knowing our past, and the lack of trust between us, there's no way this could be anything more than fun for the next few weeks.

And that's okay, because while my family is immersed in Sadie and Tom's wedding planning, I'm up for a fun distraction.

"Are you going to kiss me today?" I ask, taking a bite of pizza before internally cringing at the whine in my tone.

"It depends," he says, following it up with what I'm starting to recognize as his signature lip twitch.

"On what?" I ask.

He lifts his pant leg to display the gawdy socks I got him from the craft shop at the tree farm. "Have you been naughty or nice?"

"Oh god!" I gasp. "You weren't actually supposed to wear those."

"Really? I thought you picked them out especially for me with love and care. The Comic Sans font really brings out the holiday cheer."

I shake my head at him. "And here I thought you were a stylish dresser. I might have to revoke your fake boyfriend status."

He tugs at the hem of my cropped sweater and my core clenches. He didn't even touch me and I'm on fire.

Be careful. This is Jasper Jensen. You might get burned.

"Hmm. What can I do to get reinstated?" he asks.

"I have some ideas."

Lots of them. They've been circulating in my brain since last night.

He sweeps a section of hair behind my shoulders, giving his hand better access to my neck, and the second he touches me a shockwave of pleasure radiates down my spine.

What's it going to feel like when he *really* touches me?

He chuckles, and the sound sends a rush of warmth through my chest.

"I was thinking we'd have a friendly competition."

"When have our competitions ever been friendly?" I ask.

"You're right, I think we should declare this the first annual friendly gift-wrapping competition."

My brows reach my hairline. "You want to have a competition over gift wrapping?"

"Why not?"

"Because gift wrapping is my thing. You can't beat me."

"So this reluctance you're having isn't because you're scared to go up against me?" he teases.

"Oh, Snowflake, do you need to lie down?" I press my hand to his forehead. "It seems like you're coming down with a case of delusion."

He laughs, capturing my wrist to remove my hand from his forehead before pressing a kiss to my palm.

"All right, Sparky, let's make this contest official. We count the number of wrapped presents starting after the break. The person with the most packages wrapped wins."

I cross my arms, and push out my hip. "What do I win?"

Jasper smirks. "So confident, are we?"

"We both know I'm going to win, but I'll play along."

He leans in close to whisper in my ear. "If you win, I'll kiss you."

A kiss. Hmm.

Beggars can't be choosers, and right now, for some unexplainable reason, I'm practically begging for Jasper to kiss me again, so I'll take his kiss and then I'll make sure I leave him wanting more.

"You're on."

"May the best wrapper win," Jasper teases, then we

clean up our lunch and he escorts me back to my wrapping station before heading over to his.

For the next three hours, I wrap like I'm running a marathon.

Sandy is in awe of how fast I'm wrapping. Eventually she abandons her wrapping station, holding a water bottle in front of me to drink when I'm thirsty and a protein bar for when I need to refuel so I can keep my hands moving.

At four o'clock, we're done for the day and it's clear from the mound of presents behind my table as opposed to Jasper's much smaller one that I'm the winner.

He did give his best effort, though, so I offer a handshake like a good sport.

"Good try, Jasper."

"That's pretty impressive, Stell."

"I know." I smirk, loving that not only did I win, but my stomach is aflutter with anticipation for kissing Jasper again.

We say our goodbyes to the rest of the group and Sandy makes me promise to come back again.

We're walking through the hallway of the charity's business center when Jasper pulls me through a side door and into a small room.

"What are we doing here?" I ask, glancing around at what appears to be a meeting room. There's a small table with four chairs taking up most of the space and a picture on the wall of a group that must be volunteers or workers of the Toys for Tiny Hearts charity.

Jasper gives my wrist a gentle tug to pull me toward him.

"I've got to pay up. Give you your kiss."

Jasper presses my back against the door, then turns the lock with a click.

"Do you still want it, Stella?" he asks, his eyes searching mine.

Of course I want the kiss, it's what I asked for, but the way Jasper is intently studying my face, it feels like he's asking me about something completely different.

I swallow thickly, then nod. "Yes."

Slowly, Jasper lowers to his knees, never taking his eyes off mine.

"What—" I start to form the question but Jasper on his knees staring up at me turns my thoughts to mush. His hands hold my waist, lifting my cropped sweater until his lips have access to my stomach.

"I want to kiss you, Stella." He presses a kiss against the button of my jeans before dropping his mouth lower against the denim at my center. "Right here."

My head falls against the door, a whoosh of air leaving my lungs.

"*Jasper.*"

My voice is breathy and needy. It's a sound I'd never expected to display in front of this man. Yet, here we are. And with so much about my visit home and the low-key holiday my family is having being a disappointment, I wonder if the one thing I need right now is pleasure. In whatever form I can have it. And Jasper's tongue is offering to do the job.

"Look at me, Stella."

Years ago, at the height of our rivalry, I would have told him to fuck off. Now? Hearing the command in his voice is thrilling.

I tilt my chin down, meeting his eyes again.

"Do you want it?" His thumb brushes up the seam of my jeans and my body lights with lust and desire. "Will you let me?"

Seeing Jasper Jensen on his knees for me; there's never been a better sight.

The ache in my core is unbearable. My center is slick with need. Arousal for this man that I didn't even like three days ago, but there's also this new thing. Something I'm not sure how to label. A tenderness I've never known from him. A vulnerability in his eyes that I've never seen when we were at each other's throats, too busy puffing up our chests to outdo each other. I'm not sure how to describe what I'm feeling.

I'm sure as hell not ready to address any of it, but I am ready to see if Jasper can make me come.

Opening my mouth to speak, I stop short. Letting Jasper touch me like this will change everything. My brain is compiling a list of reasons this isn't a good idea.

This might be a joke to him. A trick up his sleeve. To play with me. To tease me. Then, leave me worked up and desperate for him.

Maybe he's trying to get my guard down, then he'll really hit me where it hurts.

But where would it hurt? He already knows that other than fake dating him, I have no romantic connections, and I'm really fucking sad that my family isn't celebrating Christmas this year because Sadie's wedding is taking priority. There's not much insecurity left to uncover.

I can be self-conscious about how he might use this interaction against me later or I can say fuck it and ride this gorgeous man's face. My body urges me onward.

"Yeah, I want it."

FOURTEEN
JASPER

I'VE MADE my success by staying focused on what I can control, then seizing opportunities when they come my way. Stella asking me to kiss her again today was a gold mine that I couldn't resist.

Yeah, I knew she'd win the wrapping contest and I'd happily oblige to kiss her.

But then I wondered if I could take things further. To show her that all I want to do is worship her.

So, I dropped to my knees in front of her and asked if I could kiss her between her thighs.

It seemed like a sweeter approach than asking to bury my face in her pussy. And she got the message because her breathing got heavier and beneath my palms, her hips started to shake.

But then, there was the longest pause. One that had me wondering if I should stand up and give her a kiss on the lips instead.

Then, she spoke.

"Yeah, I want it."

The words stopped me in my tracks because I don't

think I'd allowed myself to ever think she would agree to me touching her. Fantasies are one thing. Stella St. James giving me the green light to taste her...now I'm in shock.

"Jasper?" she whispers, and I realize I haven't moved.

"Yeah. I'm here."

"I know you are." There's a teasing lilt to her voice. "You're right in front of me."

I squeeze her hips beneath my palms.

It's the moment I'm realizing that while I'm desperate to taste her, it's pivotal. Stella St. James is giving herself to me and I better not fuck it up.

I'm afraid Stella will read my hesitation as disingenuous and call the whole thing off, but she surprises me. Her hands move to the waistband of her jeans. I watch her fingers—nails painted candy apple red to match her sweater—work the button through the hole to release it and it's a moment in my life I'll never forget.

I force myself to refocus, snapping out of whatever daze I've been in and reach for her zipper.

Hands inside her jeans, we work together to pull them down her legs. Picking up urgency, I pull off one laced snow boot, then another to free her legs from the denim.

When I come face to face with Stella's panties, I have to adjust my erection that is pressed painfully against my zipper. Red and lacy with a darker spot in the center where she's wet. *For me.*

I press my nose against the dampness and inhale her.

Fuucckkk.

The scent of her arousal has my head spinning. With my hands cupping her ass to keep her close, I kiss her there through the wet fabric.

"Jasper."

"Yeah, Stell?"

"Don't tease me."

"I'm not teasing, baby, I'm just exploring."

She takes matters into her own hands. Dipping her thumbs into the waistband of her thong, she shimmies out of it to reveal her center and a neatly trimmed patch of light-colored hair.

Eager to feel her, I slide a finger along her slit, letting her coat my knuckles with her arousal.

Exploring her further, I rub her swollen clit and massage it between my fingertips.

Stella's legs start to tremble and with how wet she is, I can gauge her orgasm is close.

I tease a finger at her entrance. "Don't come yet, Stell. I haven't even properly touched you."

A breathy laugh escapes her and she looks down on me with a narrowed gaze. "You think you're that good? One clit pinch and I'm going to come?"

I can't help but chuckle at her thinking I'm assuming her orgasm is an easy win.

I want to work for it. Every lick. Every suck and thrust. I want her to see how serious I'm taking this responsibility.

To get a better angle, I lift her right leg over my shoulder. Her hands grip my shoulders for balance, but with her hips pinned to the door, she isn't going anywhere.

And then, I have Stella like I've always wanted her. One lick through her center, and the taste of her arousal hits my tongue like I'm lapping up honey.

Now I'm the one worried about a quick orgasm. I've never come in my pants before, but if any situation would warrant it, this would be it.

Stella's hands root themselves into my hair, her fingers finding a rhythm of pressing against my scalp then tugging my hair as she uses her grip there to grind her pussy against my mouth. I play with her clit, pressing and pinching, while I fuck her sweet cunt with my tongue.

I'm going to drown on her and it's going to be my fucking pleasure.

"More, Jasper. I need more."

At her request, I thrust a finger inside her tight heat. I drag it out, then slowly push back inside her.

"You know, Stell, I've always wanted you wrapped around my finger."

I add a second finger, loving the way she clamps down around me.

"Oh, god."

"That's good, isn't it?"

Her response is incoherent, but from the way her walls are tightening around my fingers, I know she likes it. I know she's close.

"Is that what you wanted, Stell? For me to fill up this sweet cunt? Make you feel so good, you'll be begging me for more."

"I'll never beg." She sighs.

I decide to make that a personal challenge of mine.

Still fucking her with my fingers, I pull my head back to look up at her. I can't miss it. I want to see what Stella looks like when she's in the throes of pleasure. I need to see her face, especially since I'm the one making her feel this way.

"Jasper," she moans, her cheeks and neck flushing pink, her perfectly plump lips parting. Her lashes flutter and she tips her head back. Fingers gripping my hair to the point of pain, I dip back to her center for one last suck on her clit.

She cries out, her sounds unfiltered and unabashed. I love that she doesn't hold back. That it feels too good and I'm the one making her feel it.

"That's it, Stell. Soak my fingers."

She keeps rocking into me over and over as her orgasm pulses through her body. When there's no trace of it left, I press a kiss to one hip and then the other before withdrawing my fingers from her.

I lower her leg to the ground, pull her panties back into place and stand. With my eyes on hers, I suck my fingers clean, tasting every last drop of her. Her eyes widen, then dart back and forth, searching my face.

A hand moves to cover her mouth, a giddy smile pulling at her lips. "That was..."

"Remarkable? Transformative? Life altering?" I offer, but before she can speak, I pull her hand away from her mouth and kiss her hard. The kiss turns undeniably intense, so much so that I have to pull back because if I'm not careful my restraint will snap and I'll end up fucking her right here against the door.

"You okay?" I ask, tucking a strand of hair behind her ear.

"Better than okay."

I nod, then help her get dressed, tying her boots while she zips up her pants.

In the hallway, Stella glances around like she forgot where we were.

"Did I just hump your face in the Toys for Tiny Hearts office? Does that make me a bad person?"

"Not a bad person, just a horny person," I tease.

Stella attempts to playfully swat my arm, but I catch her hand and hold it in mine. She doesn't pull away so if she's freaked out about what we did, she's hiding it well.

Hand in hand, we walk down the hallway toward the exit. On our way, we run into Randall, the site manager.

"Mr. Jensen, we were delighted to have you today."

"Thank you, Randall." I nod, deciding it's a moot point

to remind him to call me Jasper again. It's been five years and countless reminders, yet he has kept the formality. "I always enjoy a day in the warehouse." I motion to Stella. "This is my girlfriend, Stella St. James."

Stella gives me a bit of side eye. Yes, we're only fake dating but if she thinks I'm not going to take advantage of that fact and introduce her as my girlfriend every chance I get, she's mistaken.

"Pleased to meet you, Miss St. James."

"You can call me Stella," she offers, and I smile at the fact that Randall will do no such thing.

Randall pushes his glasses up onto the bridge of his nose. "I'm sure you two have some big holiday plans tonight, so I'll let you go, but Mr. Jensen, I'll make sure to get you all the year-end numbers by the first of the year."

I wave him off. "It's no rush, I know you and Gloria have that vacation the first of the year, so enjoy and we'll talk mid-month."

"Thank you. We will." He turns to Stella. "And thank you for donating your wrapping skills today, Miss St. James. Mr. Jensen never told us he had such a talented girlfriend."

"Thank you, it was so much fun being a part of this."

We say our goodbyes and walk out to the parking lot.

On the way, Stella suddenly gasps and turns to me. "Oh my god, was that Randall's office? Is he going to go in there and smell sex?"

I chuckle, tucking her into the passenger seat of my SUV before climbing into the driver's seat.

"No." I turn to her. "It wasn't Randall's office. It was my office. Because this is my charity."

Her jaw drops and it's the cutest fucking thing.

"What? No way."

"Is it that hard to believe I'm a nice guy?" I pull out onto the street to start the drive home.

"No, but part of me wanted to think you were a penny-pinching CEO who doesn't care about others so I could hate you for it."

"Hmm."

"Why Toys for Tiny Hearts? And why in Cedar Hollow? It seems like access to shipping and other essentials for the charity would be better in a different location."

"The charity has multiple warehouses across the country, but I wanted to give back to the people here. Employ locals to run the warehouse and execute the vision." I pause, wondering if I should tell her the why behind Toys for Tiny Hearts. I swallow hard. "When I was younger, around third and fourth grade, there were a few tough years that my parents struggled financially. My dad's business was not doing well. They got behind on bills. We almost lost our house. At the time, Juniper and I still got a present under the tree and it wasn't until I started my company that my mom told me they had used a charity organization those two Christmases to make it special for Juni and me even when they couldn't provide it for us."

Stella blinks, then shifts her gaze to the road. "I had no idea."

"It's not something I've talked about. We were fortunate that my dad's business turned around and things got more stable financially."

"That's very personal information."

"I trust you." *And I want you to trust me.*

"Thank you for sharing it with me."

"You're welcome." I clear my throat. "Now, should we get a hot cocoa and drive around to look at Christmas lights?"

Stella's eyes illuminate with joy.

"Oh, yes. Let's do that." She claps her hands like a little kid having just been told she's going to Disney World.

"You're okay with spending the whole day with me?"

A genuine smile lights up her face. "You're not so bad."

"Good to know I'm tolerable."

"Oh, I didn't say you were tolerable, just not a bad person."

She tries to hold a serious face, but it doesn't last.

"I will say I like you more when your mouth is busy, and not talking."

"I'd agree. I like it better when my mouth is on your pussy, too."

"Jasper!" she exclaims in shock.

"What? We can do it but we can't talk about it?"

She gives me side eye, but a moment later explodes into a fit of laughter.

An hour later, when I make the turn onto our street, Stella is on a sugar high from hot cocoa and the pound of marshmallows she put in it.

"Thanks for taking me to see Christmas lights. It's one of my favorite things about this time of year."

I glance at her there in my passenger seat and hope for more days like this with her.

"It's a bummer that my parents didn't put up lights this year, but—wait a minute."

Stella leans forward to see better.

"Hold on. That's my house." She points at her house that is now decked out in Christmas lights. "What happened? How did the lights get put up? My parents must have found the time. Oh," she coos happily, "they're so pretty."

When I park the car in front of her house, she jumps out to take it all in. I round the front of the vehicle to join her on the sidewalk.

Tears in her eyes, she laughs. "I don't know why I'm crying. It's just one of those things, you know? When you're so happy and it feels like maybe things aren't as bad as you thought they were."

I shove my hands in my coat pockets, smiling at her reaction to the lights.

"I'm glad you're happy."

She clears her throat and wipes at her eyes like she's remembering who I am and doesn't want to show me her true emotions.

"Thank you for today, Jasper."

"You're welcome."

"The gift wrapping and the orgasm and the cocoa and lights viewing were the most fun I've had in a long time."

I chuckle at her list of events, and how she doesn't shy away from listing the orgasm I gave her.

"Same."

"I'm going to go inside and hang out with my family, but I'll see you on Tuesday for your family's party."

"Sounds good." I press a soft kiss to her forehead. "Goodnight, Stella."

"Night," she says, turning to go.

I'm no Grinch, but seeing the look on Stella's face when she saw the lights on her house made my heart grow three sizes. At this point, Stella has my heart, only time will tell if she chooses to keep it.

FIFTEEN
STELLA

"WHERE HAVE YOU BEEN?" Sadie asks, pushing a bowl of popcorn into my hands.

"I was with Jasper," I say, grabbing a handful of popcorn and shoving it into my mouth.

She grins wickedly. "Oh, right. Your *boyfriend.*" She winks at me. I should have never told her that we weren't really together.

"Shh." I glance around nervously. "I don't want Daniel to suspect anything."

"Daniel's out with Cady Cosgrove. They really hit it off last night and he took her to dinner and a movie tonight."

I take a moment for that to sink in. If Daniel isn't pursuing me then do Jasper and I even need to continue this dating ruse? I expect this development to make me feel relief, but that's not the emotion that comes to the surface.

"Oh, that's great." Using my upper teeth, I pull the corner of my lip into my mouth.

"What's the matter, Stell? Are you having second thoughts about rejecting Daniel?"

"No, that's not it at all. I'm happy for Daniel and Cady."

"Well, good news, it looks like you're off the hook and don't need to fake date Jasper anymore."

She has no idea how not fake the orgasm he gave me this afternoon was. It was so not fake that my legs are still wobbly from the aftershocks.

That's right. It has been confirmed. Jasper Jensen knows how to use his tongue and his fingers to please a woman. To please *me*. Never in a million years would I have thought that would happen.

But god, did I love having his mouth on me. His tongue both teasing and demanding against my clit while he thrust his fingers inside me. It was the most intense orgasm of my life. But also, the way he was with me after. Kissing me sweetly and helping me get dressed without a mention of anything in return. And from the sizable bulge in his jeans, it was clear he was turned on.

Now I'm thinking about Jasper's dick and wondering what it would feel like in my hand, in my mouth, inside… *gulp*. I try to not finish that thought, but my brain is one step ahead, already conjuring up images of Jasper bending me over that table in the warehouse office, or hiking up my skirt and drilling me on the side of my house. We really need to find more suitable places for these activities.

"Well, it's better to be safe than sorry. I should probably let this thing with Jasper play out until I'm back in New York." And collect a few more orgasms. "Just in case things with Daniel and Cady don't work out."

"Yeah, that's what I thought." Sadie's still smiling at me, her smugness evident. "You can admit you like Jasper. I won't tell anyone."

I'll admit no such thing.

"Shut up." I give her a friendly shove into the dining room where Tom is sitting at the table.

Sadie yelps out of my reach before rushing into Tom's arms. He wraps his arms around her and she playfully feeds him a handful of popcorn which he devours before snuggling into her neck.

There's an ache in my chest, and a feeling like something is missing. A longing for something that I can't quite put my finger on.

I pull off my coat and boots in the mudroom, then join them at the dining table.

Queen's "Don't Stop Me Now" is blaring from a small Bluetooth speaker on the table.

"Wow, this nineteen seventies banger is really setting the holiday mood," I say.

"We're cultivating our wedding reception playlist. Any requests?" Tom asks.

"Actually," Sadie interjects, "we're not taking requests because I already have over eight hours of music and I'm trying to whittle it down to six."

"Any Christmas music by chance?" I ask. "You know, since it's the holidays."

"Yeah, but Christmas will be over then, and everyone will be tired of hearing it, so no."

"Hey, Stell-bell," my dad says, walking in with a beer, "you want to play a game?"

"What are you playing?" I drop into the empty seat at the head of the table.

"Seating charts," my mom responds, following him in from the kitchen with a large paper layout of the reception venue and a stack of small sticky notes. "It's a puzzle and the challenge is to know all the family drama and dynamics so you don't sit people who dislike each other together."

"And my sorority sisters who are and aren't speaking to each other right now," Sadie chimes in.

I can't help the disappointment that washes over me. I thought there might actually be a family game night.

It's not that I don't want to help with Sadie's wedding to-do list, it's just I thought there would be a balance of wedding stuff and holiday traditions. It's becoming clearer that we're just going to sweep past the holidays this year and try again the next. But once Sadie and Tom are married, there will be splitting time between his family and ours. I'm all too aware that I didn't know it was happening at the time, but last year was our last time together as a family of four during the holidays.

"You'll win that every time," I tell my mom. "Your sisters are the ones that need to be flagged for violence."

My dad nods. "She's right. Janie has been on a warpath since the divorce. I'd appreciate not sitting next to her." He writes her name on a sticky note and places it at a table on the map in the back corner.

"You can't put Janie in the corner," my mom insists. "She's been through a lot. No matter how emotionally exhausting she is, she's my sister and I love her. Just like Stella and Sadie love each other."

"I love your crazy ass," I tell Sadie.

"I love yours more," she replies.

Sadie and I stick our tongues out at each other to express our sisterly love, and my mom rolls her eyes. We're grown adults now but there's something fun about acting like little children to bug our parents.

"What did you do today?" my mom asks.

"I went to Toys for Tiny Hearts with Jasper. We wrapped toys and then ended up driving around looking at Christmas lights." I toss another handful of popcorn into my mouth. "Oh, and our lights look great. Dad, you must have worked all day to put them up."

My dad thumbs through the stack of invitations, preparing to start writing names on sticky notes. "I didn't put the lights up. We were at the reception hall today with Tom and Sadie's wedding planner."

My mom sets a bowl of pretzels onto the table. "Jasper had it done."

"What?" I reach for Sadie's wine glass to wash back a popcorn kernel that has lodged itself in my throat.

"He called a few days ago and asked if it would be okay to have someone come hang everything. I told him it wasn't a big deal, we're doing a low-key Christmas this year, but he insisted and said it was a gift for you."

I'm in shock. Jasper arranged to have my family's house decorated for Christmas?

We used to compete about whose house was decorated the best for Christmas. Jasper would spend an entire weekend outside with his dad hanging lights, to put on a spectacular display while my dad pulled a Clark Griswald trying to untangle the messy strands from the previous year.

This year he could have easily rubbed in the fact that my family had done nothing to decorate, but instead he had the lights put up for us? I don't understand.

"I know you and Jasper have had your differences over the years—" my mom starts.

"Differences?" Sadie scoffs. "Is that what you call being at each other's throats for twenty years? There's been a war going on, Mom. Stella versus Jasper. Good vs. Evil. I'm surprised there hasn't been a standoff in our street with a catchy musical number for comedic relief."

"Oh, like *Westside Story*?" My mom sighs wistfully. "I've always loved that musical."

Sadie shakes her head. "Except the characters were in love and it was their families that were the issue. So that's not what's happening here."

My mom smiles. "Well, clearly you two have patched things up because you're together now."

My dad takes a drink of his beer. "I'm just happy I didn't have to climb up on the roof. If Stella's boyfriend wants to pay for light installation, I'm not going to turn it down. I've got a wedding to pay for."

"I would have helped you hang the lights if we weren't so busy with all this wedding stuff." I can tell by Tom's offering, he's feeling a bit competitive with Jasper's gift. He has nothing to worry about. Jasper's not competing for boyfriend of the year because we're not really together.

But it still makes me wonder why Jasper had the lights installed.

"I'm going to take a walk," I announce, heading for the mudroom and my coat.

"Lainey and Alex are coming over later to put together the welcome bags, you want to help?" Sadie asks.

"Yeah, I'll be back," I toss over my shoulder before heading out the front door.

On the porch, I dust off a leaf-covered chair to sit on, then take out my phone. I pause a moment, staring across the street at the Jensens' house. Then I type out a text to Jasper.

> Why didn't you tell me you had our lights installed?

He responds immediately.

> I didn't think it mattered.

Impulsively, I fire off a quick response.

> It matters.

I stare at my text wondering how Jasper will interpret it.

What about him having our lights put up is bothering me? That he did it without asking? That it cost money? Or the fact that he's my childhood nemesis and it feels like a power move?

It matters because it makes me wonder what is going on between us. I don't have a fake dating manual but I'm certain having Christmas lights installed on your fake girlfriend's house isn't in it. Neither is receiving mind-blowing orgasms but those are non-returnable so there's nothing I can do about it now.

> Please tell me you're not going to have them taken down.

> I wouldn't do that. I'm not mad you did it. I just don't understand why.

> Because I knew it would make you happy.

I stare at my phone screen, reading his text over and over.

He's right. I was happy to see my childhood home all lit up. It was comforting and nostalgic, when most of my visit home has been the opposite. But the fact that Jasper knew it would make me happy is what is getting me.

> This is very confusing. That's never been a reason for you to do something.

> It is now.

My mind wanders back to Jasper on his knees this afternoon. And in the car before that when he admitted all his sex dreams were about me. I don't know how to feel about this development. I thought I was fine with him casually going down on me today but I'm starting to question his motives. And why he didn't expect me to reciprocate. Because besides pulling his hair, I didn't even touch him. Things between Jasper and I have always been tit for tat. It feels unsettling to leave him at an advantage.

He knows what I taste like and how wet he can make me. And what I look like when I come.

Because I was horny, I let my guard down and gave Jasper ammunition. If or how he'll use it is to be determined.

Or I can level the playing field.

> I want to return the favor.

My parents already put up their Christmas lights.

> That's not the kind of favor I'm talking about.

You don't owe me anything.

> Fine. But what if I want to give you something?

...

The three dots come and go, twice, then disappear altogether. I wait a minute in case there's a delay but sure enough...nothing.

Did he just leave me on read?

An inkling of a feeling is starting to fester inside me.

Was today part of some elaborate hoax? Give me an incredible orgasm to throw me off track, decorate my parents' house, then ghost me? The adult in me wants to say fuck it, who cares? His loss and move on. But younger me, whose been on the defensive with Jasper for decades, is compelled to action.

Before I have time to think, I'm out of my chair and marching across the snow-packed street. With every crunch of the snow beneath my boots, my fury grows. Ten seconds later I'm pounding on his front door.

"Stella!" His mom, Julie, answers with a huge smile. "What a lovely surprise. I didn't know we were expecting you."

I don't have time for pleasantries, I need answers.

"Yeah, hi." I stuff my hands in my coat pockets. "I need to talk to Jasper."

"Of course. Come in. I'll grab him for you." She waves me in and the smell of gingerbread cookies envelopes me like a hug.

My mouth immediately waters.

Then, I hear it. Michael Bublé crooning "It's Beginning to Look a Lot Like Christmas."

The fireplace is roaring and their living room looks like it's straight out of a holiday décor catalog. Greenery wrapped around the banister and over the fireplace mantel. Not only the large tree in the front window, but a smaller tree I can see on the landing up the stairs. It's a tasteful explosion of plaid, red and cream. There's more holiday décor in this ten-square-foot sitting area than there is in my entire house.

"Hey." Jasper appears in the doorway. He's in the same clothes from earlier today. The sweater and jeans he was wearing when he licked my pussy, except now he's got a pair of brown loafer style slippers on his feet instead of boots. I don't know why that feels odd to me. Like I expected him to change? Because I didn't change my clothes. My underwear is still wet, or it might be wet again from the sight of his tousled hair and charming smile. Or maybe the memory of him on his knees in that green sweater is making me forget why I'm even here.

"You didn't respond to my text." My words come out sharp.

"I'm sorry. It was my turn to draw for Pictionary so I had to put my phone down."

"Pictionary?" I ask.

"Yeah, we're playing the holiday version."

Of course, Jasper and his family are having a game night. It's almost like he lured me here to rub it in my face. Except he didn't invite me, I just showed up.

"Okay, well thanks for the lights." I turn to leave because I suddenly feel out of place. Not sure what Jasper and I are doing. What the parameters are for our fake dating situation. Aware that if he wanted me to hang out with his family, he would have asked me.

"Do you want to play Pictionary with us?" he asks.

And I want to, desperately. But part of me had planned to stomp off across the street angry with him for having the Christmas experience with his family that I want.

I look longingly at the plate of gingerbread cookies on the console table. The cozy set up for an evening of old-fashioned family fun.

"I don't want to intrude."

"You're not. I want you here." He rubs my shoulders soothingly. "I thought you had plans with your family, otherwise I would have asked you earlier."

"Yeah?" These sweet confessions from Jasper are like

bread crumbs that I'm eating up. If I'm not careful, I'll end up deep in the dark forest unable to find my way out.

But fuck it, I really want to play holiday Pictionary.

"Okay." I nod. "I'll stay."

In the living room, Jasper introduces me to his aunt and uncle, his cousins and their spouses. Jasper's parents, Julie and James, give me a hug, as well as his younger sister, Juniper.

"I love your sweater," Juniper says. "You've always had great style."

"Thank you." I blush, because even though I know I have good taste, another woman acknowledging it is always nice to hear. "I love your dress."

"Thanks. I got it after I saw it on your Instagram."

"Well, it looks better on you."

Jasper's cousin, Jana, joins our conversation. "So glad you could make it, Stella. We've all heard so much about you."

"You've heard about me?" I ask, a bit shocked that Jasper would be giving so many details to his family about his fake girlfriend.

"Jasper doesn't stop talking about you. He told us all about your day at Toys for Tiny Hearts."

"I hope he didn't tell you everything." I laugh nervously.

Jasper wraps an arm around my waist and pulls me in close to whisper in my ear. "I can still taste you on my tongue."

"Stella, I might need your services in the next few days," his mom says, smiling at me.

I try to block out Jasper's words, but it's impossible, so now I'm fighting through the awkwardness of my panties being wet while his mom talks to me about gift wrap.

"Of course, I'd love to help." And I mean it, not just

because I like gift-wrapping but because I suddenly have the warmest feeling in my chest and I want to keep it there.

"What about me?" Jasper protests. "I offered to help wrap gifts."

"You'll be my backup wrapper. In case Stella's hands fall off," his mom teases.

Jasper's lips push out in a mock-pout.

"Aww, it's okay. We can't all be all-star wrappers," I tease, throwing my arms around his neck to comfort him.

"Thanks, baby." His voice is low, as if my words were truly comforting.

We stare at each other a moment, then the oddest thing happens. Jasper kisses me and it's the most natural thing in the world.

It's a soft peck on the lips. G-rated, very family-friendly. But the way my body responds is viewer discretion advised. Dirty thoughts and likely to contain nudity.

Still in Jasper's arms, he turns to the group.

"Now, everyone, I'll warn you Stella is excellent at this game."

"What are you talking about?" I ask.

"You know you're a good artist." He drops his gaze to mine and winks. "You always have been." Like how a smell or hearing a song that's tied to a memory causes it to surface, in an instant, Jasper's comment transports me back to the seventh-grade winter art fair.

Students submitted work to be purchased by donation that benefitted the food pantry. It was pretty much parents buying their kids' artwork to raise the money but it made us feel like we were contributing. I had spent hours on a winter wonderland drawing. At the time, it had been my best work and while I would be proud to show it off to the community and help raise money for the food pantry, I loved it so much I was having second thoughts about entering it.

But in the end, I wanted to help with the cause, so I submitted it to the charity art fair.

Jasper bought it.

And the next day at school, he told me with the smuggest smile on his face that he had ripped it up.

I'd fought to throw some snarky comeback at him, about how it didn't matter, or that he wasted his money, but the truth was, I'd put so much effort into that drawing, it broke my heart to know it was destroyed.

For the longest time I wasn't convinced that the fifty dollars he spent on it, that benefitted the food pantry, was worth the heartbreak of that experience. I'd have cracked open my piggy bank and made the donation myself to keep my art.

It was the moment that the feud between Jasper and I had become personal. Not simply anything you can do, I can do better. It was war.

The weight of Jasper's palm on my hip brings me back to the present.

We were kids. I should let it go. But can I?

"All right, Stella. Let's see what you got." Jasper's aunt grabs a marker from the tray at the easel and hands it to me.

I take a card from the stack, then approach the easel.

I nearly cackle when I see it says "Grandma Got Ran Over by a Reindeer." It's an elaborate one but with a few key components, the sleigh with reindeer and a detailed grandma walking down the sidewalk, my team guesses it with plenty of time to spare.

While Jasper and his team are busy guessing Jana's drawing, I steal a glance at him. I let myself strip away all the tension and hostility between us and study him objectively. The dark-framed glasses that he wears to read, those long lashes and his sculpted jaw. I'm at war with myself because the way he's made me feel in the past is colliding with the attraction and curiosity that I have to know more about him now. It's a tricky place to be and I don't like tricky. Or change. I want Jasper to be the same guy he was in high school, but that doesn't work because people change and grow. They start multi-million-dollar corporations and charity foundations. They make gestures that show you they're thinking about you and remember things you liked.

I'm more confused than ever, but I decide to take Sadie's advice and embrace what feels good.

SIXTEEN
STELLA

JASPER'S ROOM is like walking through a time portal. He still has medals and trophies from high school displayed. Band posters and dance photos.

"You don't even have to wonder what my room was like in high school. Here it is."

"I kind of like it this way. My parents cleared my stuff out and now it's got the generic appearance of guest room number one."

I lean against the desk, taking in his room. Trying to imagine him doing homework here or talking on the phone with girls.

I glance at his bed. Hooking up with girls.

"You said you had something you wanted to give me? Or did you change your mind when my team beat yours at Pictionary?"

"Hey, I did what I could. Your Uncle Ron was our weak link."

His lips twitch with amusement. "Yeah, he drinks too much eggnog to be helpful."

"So you sabotaged my chances of winning by putting me on his team?"

"Truthfully, I thought your skills would overcome his weakness, but alas it was not enough."

My mouth gapes open in mock outrage. "You're not doing yourself any favors, you know."

It's what I say, but I walk over to where he's sitting on the bed. The moment I'm close enough, his hands move to the back of my legs, his warm, firm grip sending a rush of desire between my thighs.

"Stella St. James in my childhood room. What an incredible sight."

"Hmm." I push my fingers through his hair before removing his glasses and setting them on the nightstand. "What does your bedroom look like in LA?"

"It's boring. Wood bedframe, gray bedding." A teasing smile pulls at his lips. "A photo of my fake girlfriend on the nightstand."

"That would be weird, seeing as we only started fake dating three days ago."

His smile doesn't waver, which makes me curious what his motives are here.

"Why are you single, Jasper?"

"Are you inferring I'm quite the catch and a woman should have snapped me up?"

"I'd never give you the satisfaction."

"Because I spent my early twenties developing and patenting the proprietary software at Jensen Innovations, and lately I've been working eighty-hour weeks trying to get our New York office set up."

I trace my finger along his jaw. It's something I realize I've been wanting to do since I laid my head in his lap on the airplane.

"Why are you single, Stella?" he asks.

"Because I can't find a guy that can handle me."

Jasper's hand moves from my hip to slide under my sweater. His fingers splay over my ribcage, the tips teasing at the band of my bra.

"I think I handled you just fine this afternoon."

I can't help the goofy smile that pulls across my face. "Yeah, you did."

I lower my hands to grip the hem of my sweater and tug it over my head, then toss it onto the bed behind him. Then, I unclasp my bra and do the same.

Jasper blows out a heavy breath, and the air tickles my chest, causing my nipples to stiffen.

"God damn, Stella. You're incredible."

His large hands work their way up my ribcage, his thumbs caressing the underside of my breasts.

He pulls me into his lap, his mouth latching onto one hardened peak. My center immediately floods with arousal at how good it feels to have his mouth on me.

I grind down onto his erection and let myself bask in the euphoric feeling of his strong arms wrapped around me. His hold makes me feel like a precious thing he needs to protect. And then there's his glorious mouth worshipping my breasts. *Damn.* He makes me feel so good.

"It's the craziest thing. I look at you and I know you're the guy I've fought with for so long, but for some reason the wires in my brain get crossed and I want to rip your clothes off and find out how big your dick is." I groan when he sucks my other nipple into his mouth. "I'm sure it's perfect like everything else."

I swallow, my eyes dropping to the bulge beneath Jasper's zipper.

Yup, Jasper's dick is going to be a good one. I can already tell.

He cups my face with his hands and kisses me like I'm his. He tastes like gingerbread and cinnamon. It's the same kiss that had me surprised last night at the bar, and again today in the charity office. A kiss that has no business in this arrangement, yet I can't get enough.

Jasper's embedded himself into my brain, and I plan to do the same to him.

I extract myself from his mouth and sink to the floor between his legs.

"Stella," he groans, reading my intention.

"Hmm?" My hands work quickly to undo his belt, then button and zipper.

"I never said thank you for having my family's Christmas lights installed."

"Hey," Jasper lifts my chin with his fingers. "I meant what I said. I did it because I knew it would make you happy. No other reason."

I believe him. Or I want to. Either way, I need to see him come undone.

"Now it's my turn."

I palm his erection over his boxer briefs, my core clenching at feeling his thick shaft. "I want to feel you in my mouth, Jasper."

"And I want to fuck this pretty little mouth of yours, Stell." He caresses my cheek, his thumb teasing the bottom of my lip before he leans back to assist me in pulling his jeans and boxers down his legs.

For a moment, I sit back on my heels and stare. Jasper's cock is the biggest I've ever seen. It's thick and long and the hair at the base is neatly trimmed.

It's perfection. Normally I'd be annoyed with him, like

he planned to dangle this perfect cock right in front of my face, but I'm the one who insisted I suck it so now I've got to swallow my pride and his cum.

It's like I've momentarily forgotten everything I know about sucking a dick. Like I've never done it before. All the dicks before Jasper's seem to be wiped from my memory.

He's watching me intently so I start with the basics.

I lick him from base to tip.

"You're going to need more than that, Stell."

I know he's right. If I try to lick him like a lollipop, I'll be here all night.

With a mischievous glint in his eyes, and a cocky smirk on his face, he tells me, "Spit on it, baby."

Those four words do something to me. I'm not sure if it's the first three or the last one, but I do what he says. I collect as much saliva from my mouth as I can and leaning over his lap, I let it drip from my lips and onto his shaft. With my hand, I work the saliva up and down, coating him thoroughly before diving back in with my mouth. Swirling my way down his shaft with my tongue, I suck him in as deep as I can take him.

"Bet you never thought you'd be on your knees with those pretty lips wrapped around me."

"Never." I moan around him. I'm so wet, I know I'll have to go home and touch myself when I'm done here. I'll have to muffle my moans under my pillow so I can make eye contact with my family tomorrow at the breakfast table.

He groans. "Fuck, Stella. You suck me so well."

I love hearing him say my name. Knowing that I can draw out as much pleasure from him as he wrenched from me earlier.

"Are you going to be loud?" My eyes look to his

bedroom door while I continue to pump him with my hand. "Do I need to stuff my panties in your mouth?"

He smirks, a wicked gleam in his eyes. "That's an excellent idea."

In the quickest power move I've ever seen, Jasper reaches for my hips and flips me around, lying back on the bed with me on top of him so my center is aligned with his face. I'm face down, still holding onto his hard cock but now I can feel his warm breath at the crotch of my very wet panties.

I'm so shocked by the change in position, I can't even form words. "W—what are you—"

"You think I'm going to pass up a chance to taste you, baby?"

His tongue plays with me through the soaked fabric. Then, in a torturous turn of events, he pulls my panties aside and dips his tongue inside me.

"You going to finish me off, Stell?" he asks between licks. "Or are you too distracted now?"

I suck him to the back of my throat and he groans against my center. "What was that? I couldn't hear you with all that groaning."

He thrusts a finger inside me, then another. A third joins and oh god, I'm going to lose it.

"You should see how hot your pussy looks filled with my fingers."

I bite back my own groan, every nerve in my body tingling. It knows how good this is going to be, but I can't give in yet.

"Damn it, Jasper," I mutter before picking up the pace on his cock again.

He adjusts my hips, lifting me closer to him, then starts a punishing rhythm with his tongue against my clit.

"I bet I can make you squirt on my face," he mumbles against me.

"I bet I can suck you dry," I counter before returning to the rhythm that I know is pushing him closer to the edge.

It's the most satisfying sexual moment of my life. Feeling the pleasure building inside me while holding it off so I can make Jasper eat his words and blow his load down my throat first.

And it pays off.

"Stella, baby, I'm going to come."

It's a warning that almost has me smiling around his cock, but I refrain because I've got to take him to the finish line.

And I do.

"Uhhh, fuck. *Stella*." He groans against my pussy.

My name on his lips at the peak of pleasure is something I'll never forget. It's the satisfaction of every victory over Jasper combined.

A moment later, his cock jolts beneath my palm and the hot liquid of his orgasm hits the back of my throat. Even through his orgasm, he keeps fucking me with his tongue. Once I know I've won, I let myself focus on my release.

"Now come all over my face, Stell. I want it messy."

With my hands on his thighs, I ride his tongue until my body coils tightly. Then, with my pussy full of his fingers stroking me in that perfect spot, and his lips suctioned around my clit, I splinter into a million pieces. A rush of wetness escapes from my center and I not only hear but feel the satisfied rumble coming from Jasper's chest beneath me as he laps up my orgasm with his tongue.

His warm hands massage along the backs of my thighs and over my ass. For the second time today, I find myself relishing in Jasper's touch and attention. But I can't stay here forever. God, I wonder if his family suspects what we're doing up here.

I roll off him and collapse onto a heap on his bed.

"Come here," he says, pulling me into him.

"I'm sweaty," I pant, trying to get my breathing under control.

"I like you exactly like this." His fingers splay over my ribcage, his thumb teasing under the weight of my breast.

I yawn and curl into him.

"You can stay," he offers.

"I better not. I'm supposed to help Sadie pack welcome bags for the hotel guests."

He nods, then finds my underwear while I put my bra back on. There's something surreal about putting your clothes back on after they were thrown off in a moment of passion. I'd expect it to be awkward with Jasper because what we are to each other only exists in the tiniest overlap of the fake dating, not friends but suddenly we have benefits, Venn diagram.

He walks me downstairs and I'm thankful his family is nowhere in sight.

"You good?" he asks, brushing my hair over my shoulder before he helps me put my coat on.

"Of course I'm good. I'm just tired."

He pulls on his shoes and walks me across the street, which is silly and I tell him so.

"Thanks for the light installation. And the orgasms." I try to make light of the fact that this thing between us has gone from zero to sixty in less than a day, and I'm not exactly sure what to make of it. Now that I think of it, other than our friendly wrapping competition, which now seems like an excuse for Jasper to put his head between my thighs, we haven't fought all day. The realization has me gasping with surprise.

Beneath the porch light, Jasper's eyes light up knowingly.

"Don't tell me we're becoming friends, Stella. Next thing you know you'll be saying you actually like me."

"Never," I say, but it comes out breathless, instead of resolute.

"Stella?" he calls right before I open the door to my house.

"Yeah, Snowflake?" I tease.

"Sweet dreams." He winks.

SEVENTEEN
JASPER

I STARE at my computer screen and will myself to focus on the memo for next year's company goals and key performance indicators, but memories of two nights ago with Stella have me in a chokehold.

It was one thing for her to let me taste her at the warehouse, but when she dropped to her knees for me and sucked my cock to the back of her throat, I realized that if I wasn't already in love with her before, I am now. It's been two days since she played Pictionary with my family then gave me the hottest blow job I've ever had. She's been busy with appointments for Sadie's wedding, and I'm dying to see her at my family's Christmas Eve party tonight.

The doorbell rings and I move to stand, but then I hear Juniper yell, "I got it."

I'm assuming it's one of her friends, but then she appears in the doorway, and I do a double-take when I see my best friend, Liam, following closely behind her.

"Bro, why didn't you tell me you were coming?" I ask, standing to greet him with warm hug.

"Wasn't sure of my plans until this morning when I got

on the plane. I hope the invitation still stands or it's going to get a bit uncomfortable."

"Of course it stands. I'm just surprised you showed."

"You said tonight was going to be a rager and you know I can't resist a good party."

"It's got nothing on Elton's party," I tease, knowing full well Liam has invitations to many celebrity parties in London this week, "but we'll try to show you a good time.

"You remember my sister, Juniper." I motion to where she's leaning against the doorway watching us.

Liam gives her a quick nod and a smile. "Of course, I remember June Bug."

Behind him, her face scrunches at the nickname that I've affectionately used for years, but she recently told me to stop.

"She's opening a bookstore soon."

"A romance bookstore," Juniper speaks up. "I just signed the lease for the space last week."

Liam nods at Juni. "Congratulations. It must be in the Jensen blood to be an entrepreneur."

I chuckle. "Yeah, but I told her I'm shit with the business side of things, and she should talk to you."

"It's fine." Juniper throws out a dismissive wave. "I'm sure you've got a million other things to do."

"Never too busy for Jas's little sis. Let's find some time to chat tonight."

My mom walks into the living room and gasps with excitement at Liam's appearance. "Oh, wow. You flew all the way from London for our Christmas Eve party? We're so honored."

"I didn't make it home to London. I was in Vancouver visiting another friend, then thought I'd stop here."

"We're very excited to have you." My mom hugs Liam. "We'll put you in the basement guest room."

"I got a room at the Snowshoe Inn. I didn't want to impose."

"Nonsense. You'll stay with us."

My mom takes his coat and moves toward the closet, when the doorbell rings again.

"Oh, that must be Stella," she says, hanging up Liam's coat.

I check my phone for the time and to see if I have any missed texts from Stella. "I didn't know she was coming over now," I say, following my mom to the front door.

"She's coming over to help me wrap a few gifts." My mom turns to Liam. "Stella is an excellent gift wrapper. She used to volunteer at the outlet mall every holiday season for their free gift wrap program."

Liam nods in understanding.

I smile at the memory. I used to take gifts to Stella to wrap. Mainly so I had a reason to talk to her but also to challenge her superb wrapping skills with oddly-shaped boxes. Some of the items weren't even presents, but stuff we had around the house.

"Merry Christmas Eve." My mom greets Stella with a warm embrace.

"Merry Christmas Eve!" Stella greets her with equal enthusiasm.

Stella's wearing black leggings with an oversized cream sweater with a red scarf wrapped around her neck. My mom takes Stella's coat and I hover nearby while Stella removes her lace-up snow boots with faux fur trim.

"Hi." A grin springs across her face.

"Hi." I pull her in close, one hand affectionately cupping her ass, before brushing my lips against hers. She melts against me and it's the most satisfying feeling. Where we were playing at being a couple before, and failing, after the day at Toys for Tiny Hearts something has shifted between us and I'm soaking in every moment.

"So this is the famous Stella." A wolfish grin slides onto Liam's face as he extends his hand to her. "Liam Hargrove."

"Liam's my friend and business partner at Jensen Innovations," I fill in for Stella.

"It's nice to meet you," she says, taking his hand. "And why am I famous?"

"You're the woman that has finally taken this bachelor off the market."

Stella's mouth twitches. "Well Jasper has been begging me to go out with him, and he finally wore me down."

Liam chuckles. "That sounds right. I've always known him to go after what he wants."

"It's the sweetest thing that they're dating now. They used to be rivals in school and downright mean to each other, but if you ask me, all that fighting was foreplay."

"Mom," I groan. I can typically brush off my mom's antics, but right now, this isn't the conversation I want to be having in front of Stella. "Please don't say foreplay."

"I used to tell Jasper, it was only a matter of time before Stella returned his feelings."

For her part, Stella doesn't let on. She wraps her arms around my waist and adoringly looks into my eyes.

"It was only a matter of time before we found each other, right, Snowflake?"

I wrap my arms around her and kiss the top of her head. "That's right, Sparky."

EIGHTEEN
STELLA

THE JENSENS' Christmas Eve party is a tradition in Cedar Hollow. I went a few times when my parents forced me to but once they realized how treacherous it was to have me and Jasper in the same room, they stopped making it a requirement. I'd heard from friends over the years about how fun and festive the party was but could never bring myself to show up on Jasper's turf.

But tonight, I'm attending the party as Jasper's date as my part in our fake dating arrangement. He's held up his end of the bargain, so now I'm here to keep his mom from meddling in his personal life by playing matchmaker.

While I'm normally a cozy sweater and jeans girl, I also love to dress up. I'm especially excited to wear the pine green velvet dress with black bows gathering the front at my neck and cleavage. It's sophisticated, yet sexy, and I'm dying to see Jasper's face when he sees me in it. I love the way his eyes light up and the skin around his eyes crinkles because his smile fills up his face.

The moment the thought is out of my head, I stutter step in my black heeled booties.

Let me try that again.

What I mean is, I want Jasper to see me in this dress and fall all over himself because it'll be payback for all the times he thought he bested me when we were growing up.

Yeah, that feels better. I think.

I shake the thought loose, and pull on my coat to make the short walk across the street.

Jasper asked if he could pick me up, but after helping his mom wrap some gifts, I was in a rush to shower and change so I told him I'd meet him there. Besides, his friend and business partner, Liam, is here now and I wanted to give them some time to hang out.

If I thought the Jensens' house was magical a few nights ago, the party preparation has turned it into a full-on Christmas wonderland.

Twinkling lights are wrapped around the banisters and over the windows. Pine garlands and wreaths add to the décor but also give a forest-fresh scent. A few high-top tables in the foyer are draped in white linen with pine branches, candles, and ornament centerpieces.

It's a Wonderful Life is playing on the framed television over the fireplace, while a mix of holiday classics makes up the soundtrack for the evening. The food table is a work of art. Mini meatballs, puff pastry bites, and cheeseboards are laid out on a table for snacking with rosemary and cranberries for garnish giving the table a festive vibe.

I'm tempted to take pictures just so I can remember every detail because right now it feels impossible to take it all in.

There's a crowd of people at the bar in the foyer that's serving a variety of spirits like mulled wine, spiked eggnog, and peppermint cocktails, so I hang my coat on the designated rack in the hallway and walk into the living room.

I spot Juniper putting the finishing touches on the hot cocoa bar and dessert table so I walk over to help her.

"You look stunning," I say, taking in her black sequin skirt and black silk blouse before embracing her.

She stands back to admire my dress. "Thanks, you do, too."

"Do you have a date tonight?" I ask, glancing around.

She shakes her head. "No, I'm keeping my options open."

"Anyone you want to get caught under the mistletoe with?" I tease.

She scans the room, and I swear her eyes linger on Liam, Jasper's business partner, who's talking in a small group across the room, but then she quickly directs her eyes away.

"Potentially."

I decide not to pry because that's all anyone has been doing with Jasper's and my relationship, and I don't want to be that person.

"Jasper told me about your romance bookstore. That's very exciting. Congratulations!"

"Thank you. There's still a lot of work to be done, but hopefully I'll be open by next summer."

"Well, I can't wait to see it the next time I'm in town. Oh, and my best friend is a romance author."

"What's her name?" she asks.

"Pippa Monroe."

Juniper's jaw drops. "Are you serious? She's like my favorite author."

My smile widens at that. "She'll have to come visit your store then."

"Oh my god, I would die."

"Don't do that, then you won't get any signed books."

"Right." She laughs. "Tell her I'd love to have her visit anytime."

"I'll make sure she knows."

I scan the room of guests, suddenly wondering who the woman that Jasper's mom had intended to set him up with is. I know she won't do it now that she thinks we're together, but I can't help but be curious who Julie Jensen thought would be a good match for her son.

I turn to Juniper. "Hey, do you know who your mom was going to set Jasper up with?"

Juniper blinks in confusion. "Huh?"

"Before she found out we were dating, of course."

I don't really think this other woman is competition, but I also wonder what will happen when Jasper and I are no longer fake dating?

"I guess she lives in New York, too, and since Jasper is moving there, she had thought they would hit it off. That's before she knew we're together of course. It's a thing your mom has done at this party. Tried to set Jasper up with a date."

Juniper presses her lips together. "My mom wouldn't do that."

I'm surprised by her statement. "Why not?"

Her mouth pulls into a huge grin. "She knows Jasper wouldn't be interested."

"Because he works so much and has no time for a relationship?" That's the reason he told me he was single and hasn't been dating.

"Sure. That, too."

Jasper's arms wrap around my front, one large hand settling on my stomach causing a swarm of butterflies to take flight.

"There's my girl."

"Hmm." I let myself indulge in his embrace for a moment before I cut it short and turn to find him looking ridiculously handsome in slacks and a wool blazer, with a white button-down underneath. He's wearing his glasses which I've decided are my weakness. His hair is perfectly styled in that messy wave I like and his jawline is clean shaven, making every angle perfectly cut.

He's beautiful. He's charming. And, he's in trouble.

"You look absolutely stunning." He pulls me in a for sweet kiss. "Does this dress wrinkle?" he asks huskily against my ear.

But I can't be distracted by a sexy, nerdy tech guy that fills out his slacks just right.

"A word, Jasper." I give him my best admonishing tone before reaching for his hand and pulling him from the group. I march him straight upstairs to his room where I know we'll have privacy to talk. But we'll do nothing else, because I'm upset with him.

When we're in the quiet of his room, Jasper turns that magnetic smile of his on me.

"You're so beautiful." He reaches for a strand of my hair, but I gently capture his wrist and guide his arm down.

"No one can even hear us right now."

He shrugs. "Doesn't matter. I'm being honest."

I give him a stern look and cross my arms over my chest. "Didn't you tell me your mom usually sets you up at this party? That was the reason you needed me to be your fake girlfriend."

"Okay."

"That's not an answer. I just talked to Juniper. She says your mom never sets you up. That your mom knows you wouldn't be interested so it's pointless. So why did you tell me that?"

He sighs. "Juniper's right. But my mom did set me up once and it could have happened again."

"That's your basis for needing a fake girlfriend?" I scoff.

"No. I knew you needed my help but you wouldn't accept it if you knew it was one-sided."

"You're darn right." I nod, my frustration with Jasper growing. "So, you lied and didn't need me at all?"

I'm surprised by the hurt in my voice. It's not even his lie that I'm having issue with, it's that this entire time, he's been helping my cause knowing there was no need for reciprocation. It tips the scales in his favor.

"That's not completely accurate. I stretched the truth, and I did need you." He pins me against the door. "I still need you."

"What for?" I search his face, my breath growing heavy and ragged.

"Because you make the holidays feel like something worth celebrating. Spending time with you this past week has been the most fun I've had in forever."

His admission is surprising and it makes me feel like less of a charity case. But I'm not sure how to act now that Daniel's distracted with Cady and Jasper doesn't even need a fake girlfriend.

"We should get back to the party," I tell him.

Jasper steps back to give me space, but instead of following me out the door, he's across the room grabbing something off his desk.

"Stella, wait. I want to give you your Christmas gift."

I stare at the black square velvet box he's presenting me. It has the largest red bow on it. It's completely disproportioned to the box, but I also kind of love it for that reason.

"I didn't know we were exchanging gifts." My voice is soft.

"We never said either way. I saw these and thought of you."

I feel his eyes on me as I untie the ribbon, then carefully open the box. Sitting in the box are the most exquisite pair of bow-shaped diamond earrings. They're large enough to be a statement, but small enough to wear every day. They're perfect.

"*Jasper,*" I gasp.

"Do you like them?"

I look up to see him watching me intently.

"I *love* them."

He presses a kiss to my lips. "Good. I know they'll look great on you."

I'm confused. And empty-handed. I didn't get Jasper a Christmas gift. With all the chaos of our fake relationship and Sadie's wedding preparation, I hadn't thought about it. That only makes me feel worse, because Jasper thought about me.

"I didn't get you a gift."

"This is all I want for Christmas, Stell." He slides a hand over my ass and pulls me closer. "You, here with me. Is that okay?"

"Yeah." I nod, simply because I'm under some kind of spell.

There's something about the way he cups my ass. He did it earlier when I came over to help his mom wrap presents. It's possessive, yet sweet. The perfect combination of firm fingertips pressing into the flesh of my ass, and his palm gently holding me to him.

He leans down to give me the sweetest, yet most passionate kiss. I kiss him back with everything I have, hoping that fills the void of not having a Christmas gift for him.

I might take this chance to maul him, but I want to spend time with our families and friends, so I exchange my current earrings for the diamond bows Jasper gave me and we head back downstairs.

The evening is the most fun I've had in a long time.

Liam is a good time. Telling jokes—I love his British humor—and whipping up the perfect dirty martini, which I only need one of so I won't faceplant into the Christmas tree.

Jasper and I end the night slow dancing in the living room to Frank Sinatra's "Have Yourself a Merry Little Christmas," then I help his family with some clean up, before he walks me home.

It's snowing big fluffy flakes, so I stick my tongue out and capture one.

I turn to find Jasper staring at me.

At my front door, he pulls me in close and kisses me as the snow falls around us. Everything about tonight feels like a scene out of a Hallmark holiday movie, except for the part where we slipped into the bathroom together so he could finger me while I gave him a skilled hand job.

"Merry Christmas, Stella." He kisses me softly.

"Merry Christmas, Jasper." I kiss him back. He tastes like citrus and whiskey and gingerbread. I'm tempted to invite him up but not ready to analyze what that means for our fake relationship status, so I say goodnight.

In my bedroom, I get ready for bed, but I leave the curtains open so I can enjoy the view of Jasper's house, then crawl into bed and fall asleep.

❄

Christmas day with my family is magical. There's nothing that can be done for the wedding today. No vendors to call or meet with, no shops open to run errands at, so we lounge in the matching pajamas that I got for us—except Daniel, since I didn't know he was going to be here—open gifts, then stuff ourselves with a home-cooked meal that we all helped prepare.

We clean up the kitchen, play a few rounds of Spades at my dad's request, then gather up snacks to watch a movie.

It's a wonderful Christmas, exactly what I wanted, but something is missing.

Not something. Someone.

I miss Jasper.

There. I said it.

"Is Jasper coming over?" Sadie asks like she can read my mind.

"No, he's got plans with his family. And Liam." I don't know why I don't tell Sadie the truth. The truth is, I don't know what his plans are. She knows we're not really together, so I'm not sure why she would expect him to show up today.

"Are you sure about that?" she asks, pointing out the window where Jasper is crossing the street to our house right now.

The doorbell rings and I race for it. Throwing the door open, I leap into his arms, which doesn't bode well for the wine bottle he's carrying, but Jasper performed juggling in the sixth-grade talent show, and he manages to catch me and not drop the wine.

I maul him in front of my family. It's what I would do if we were really together and until after Sadie and Tom's wedding, the show must go on.

Out of breath from our make-out session, he pulls back with a huge grin on his face. "Merry Christmas, Sparky."

"Merry Christmas, Snowflake."

"Are you guys going to watch the movie?" Sadie yells as we pass the living room on the way to the kitchen.

"Yeah, but you can start it without us."

I lead Jasper into the kitchen under the pretense of getting snacks and drinks for the movie, but I'm really luring him there to continue our make-out on the kitchen counter.

"What are we watching?" he asks.

"*Elf.*"

"Classic."

"Want to know something weird?" I ask.

"Always."

"I missed you today."

The corner of his mouth crooks up in a satisfied smile.

"Don't be smug about it."

"This isn't smug. This is happy." He presses a kiss to my lips. It's innocent and sweet, and I want more.

I realize it in that moment, I'm happy, too. Happy that he's here. Happy that he's made being home for Christmas as magical as I hoped it would be.

"Will you come somewhere with me tomorrow?" he asks.

Anywhere.

"Where?"

"A quiet mountain cabin. Just you and me."

"Is there a bed?" I ask. He already had me at 'just you and me' but I need to do my due diligence.

"Yes."

"Are we naked?"

"If we want to be."

I vigorously nod my head. "We want to be."

"This feels like another gift, when I haven't even given you anything yet."

I'm tempted to make him come upstairs with me right now, but with everyone in the house watching a movie, it's too risky.

"Trust me, Stell. You're going to be giving and receiving plenty this weekend, so make sure you get adequate rest and hydrate."

I smile at that because for the first time, Jasper and I are on the same page.

We join my family in the living room, snuggling together on the couch. Jasper sits behind me, and I'm in his lap with my head tucked under his chin while we share a bowl of buttery popcorn. When Jasper can't reach the bowl, I pinch a few pieces of popcorn in my fingers and reach it over my shoulder to his mouth. He eats the popcorn, then teasingly licks my fingers.

I glance around to see if anyone is observing how adorable we are, but everyone's eyes are on the movie. They're missing a good show.

Look at this performance. We're naturals.

And it does feel natural with Jasper.

That inkling from last night rears its head again. If no one is watching, what's the point in pretending? And are we even pretending anymore?

NINETEEN
JASPER

"NOOOOO! You've got to be kidding me!" Stella yells from behind the bathroom door.

I pause, mid-pull on the cork from the wine bottle I'm opening. We just arrived at the cabin I booked for the next two days. A small window of time between Christmas with our families and Stella needing to be back for wedding festivities.

We've both been craving time alone that wasn't behind a locked door, afraid one of our family members would knock or in Stella's case, hear her muffled screams.

"Everything okay, Stell?" I call.

The toilet flushes and I hear the faucet running. A moment later, the door swings open and Stella comes flying out.

"Well, the fuck-fest is off."

I chuckle, brows lifting. "Fuck-fest? I wasn't aware the weekend had a theme."

With the cork now free, I pour us each a half glass of pinot noir.

"You know as well as I do that's why we're here." She glances into the living room where the fire I started is roaring. "Look at that fire, and the snow falling outside. And you have wine." She takes the glass from me, her gaze dropping to my outstretched hand. "Even the veins in your hands are giving off fuck me vibes," she cries.

I'm not sure what happened in that bathroom, and part of me is afraid to ask.

"I think I missed something. Why are you upset?"

"I just got my period." She groans, taking a hearty gulp of wine. "I mean I knew it was going to happen soon, like in a day or two, but apparently my body didn't get the memo that multiple orgasms were to be had this weekend. Couldn't it have waited? It feels like sabotage."

She sets the wine glass down and claps her hands. "Oh, I know! We can dry hump while you play with my tits. It's not what we had planned, but it's something. And I'll definitely be giving you a blow job because this place," she motions around to the luxury cabin before turning back to me with both hands thumb to index finger and the remaining three fingers upright, "is NICE."

"Stella," I wrap her up in my arms, "I want to spend time with you. Naked time was going to be a bonus."

"You're just saying that so I won't fall into an emotional pit of despair."

"I mean it." I release her so I can show her the bag of books and games I packed. "See? I wouldn't have brought Holiday Monopoly if I thought we'd only be having sex. That's at least a three-hour commitment."

"Well, you're clearly not as horny as I am. I only packed lingerie and sex toys." She sighs. "I'm thinking the crotchless panties are not going to be useful anymore."

"Stell, the things that come out of your mouth never cease to surprise me."

She takes a sip of her wine, then smirks. "If you play your cards right, one of them could be your dick."

I press a soft kiss to her mouth.

"Why don't you go change into some comfy clothes and I'll order us in some dinner. We'll watch a movie, eat chocolate, and I'll give you a foot massage."

"You know you don't have to do the whole fake boyfriend thing while we're here. The only person I'll tell about this is Sadie and she already knows we're not really together so..."

"I'm not your fake boyfriend right now, Stell. I'm just a guy trying to make you more comfortable when you're not feeling your best."

She nods, considering this a moment while she examines the Monopoly box. "I get to be the banker, right?"

I chuckle at how quickly she got on board with the game. "The job's all yours."

"Okay." She turns to go. "Wait. I really don't have any other clothes but these," she points to her jeans and sweater, "and lingerie."

Now it's my turn to smirk. "Check the closet."

Her eyes narrow at me suspiciously, but she disappears into the bedroom.

A few minutes later, she returns wearing the burgundy-colored cashmere pant and sweater lounge set I brought for her.

"You have to stop buying me things," she protests, but I can see how much she likes the outfit, and it looks perfect on her.

"I like buying you things."

"Okay, well now that you mention it, there is something else that I need."

"What's that?" I ask, knowing already that I'll get her whatever she wants.

"Tampons." She smiles teasingly.

I reach for my keys. "I'll be right back."

"What movie are we watching?" I ask, spreading the food out in front of us.

"*The Holiday,* of course."

The cabin I rented is a twenty-minute drive from Cedar Hollow, so I made the drive to the store to get Stella's tampons, and picked up dinner from Flores Mexican restaurant, which I know Stella loves, on my way back.

I spread out our feast. Tacos, nachos, and their famous tamales.

"Which one is that?" I ask.

"Cameron Diaz, Kate Winslet. They switch homes for the holidays and end up falling in love with guys at their new locations. And Jude Law is gorgeous, especially when he wears glasses. It's only one of the best holiday movies of all time."

"Okay, let's watch it."

"Wait. You've never seen it?"

"I'm sure I've seen parts of it. It sounds familiar."

"This is my favorite holiday movie. You're going to love it."

We start the movie, then dive into the food. Twenty minutes in, it's clear that I won't actually be watching the movie, because watching Stella watch the movie is more fun. She's giggling and kicking her feet like she's never seen it before. Me, I'm just enjoying being close to her.

I hit pause on the remote. "Hold on. The asshole's name is Jasper?"

Stella presses her lips together trying not to laugh, then shrugs. "I mean, if the name fits."

She gets pinned down and a twenty-minute make-out session for her impertinence before we can continue watching the movie.

When the movie is over, Stella turns to me. "Okay, favorite part?"

"I did enjoy seeing Cameron Diaz in that lacy little bra thing."

She gives me a playful whack with the pillow. "That's it?"

"Okay, I got it. The moment her character, what was her name?"

"Amanda."

"Yeah, Amanda. When she was driving off in the car, and realized she didn't want to leave him like that. She let down her guard and took a chance."

Stella smiles in approval.

"What's your favorite part?" I ask.

"When Amanda and Graham are lying in his girls' fort and it's got all the twinkle lights and cozy blankets and pillows. And they look at each other and just know that there's something between them."

Her smile is radiant. And genuine.

I love Stella's smile. I've seen it from afar, but I've never been in the direct line of it like I have the last few days. It's like basking in the warmth of the sun on a perfect spring day after a long winter in the shadows. It's addictive and I want more.

I want Stella to be mine. For real. It's what I've always wanted.

It takes some effort and some imagination, but we build our own makeshift fort in the living room near the fireplace, then pull the pillows and comforter off the bed and set them up inside.

She slips her hand under the hem of my shirt. My abdominals contract under her fingers' exploration. Her touch is tender and playful, nothing like how urgent and rushed most of our intimate time has been.

"Where are you going to live in New York?" she asks.

"I haven't found a place yet. Do you have a recommendation?"

"I live in Chelsea. There are some great places in my neighborhood. I guess it depends on what you want to be close to." She pushes farther up my shirt, finding my nipple and rubbing her thumb across its surface. "Like your office, or things you enjoy like going to a particular gym or being close to the High Line if you like to run."

I nod, loving the feel of her hands on me. "All good things to consider."

"Remember how you said there were millions of people in New York City and we'd probably never run into each other?" she asks.

"Yeah." I had said it to make her feel better about me moving there, not because I wanted it to be the truth.

"What if we wanted to?" Her eyes lift to mine and I see that same vulnerability in them that was there when she was sick on the plane with her head in my lap.

"What are you saying, Stell?" I nudge. I need her to give me something more.

"I like you, Jasper," she whispers. I can't contain my smile.

"I like you, Stella."

"So maybe we can like each other in New York, too."

"I'd like that."

"There's so much liking going on between us. I don't know how to deal with it," she confesses.

I know this is harder for her. I've loved her for years and this is all new to her.

"We can figure it out together."

With the fire roaring beside us, we kiss for hours. Exploring each other slowly, teasingly. Stella strokes me while I slip my fingers into her panties and play with her clit. We come apart together, then I hold her all night under the canopy of our makeshift tent.

The next day is more of the same.

We play Holiday Monopoly and Stella wins, but she gives me a conciliatory blow job that makes me feel like I'm the real winner.

After we decorate the gingerbread houses I had delivered, we find a sledding hill nearby where we use the sleds provided at the cabin. I crash into a tree, breaking one sled. Stella's concerned I hurt myself and even though I'm fine, I let her fawn over me. Then, we share her sled for the rest of our time on the hill. I take a million selfies with her and save one to my phone's lock screen.

We cook dinner together. Spaghetti and meatballs with salad and garlic bread. We open another bottle of wine, and Stella dresses in her finest lingerie but ends up stealing my sweatshirt because she's cold.

It's simple and domesticated, and I want this with Stella forever.

We leave the cabin Saturday afternoon and when I drop her at her house, she wraps her legs around my waist and kisses me like I'm the oxygen she needs to breathe. No one is there to witness it.

"I'll be in wedding planning jail for the next few days," she says, kissing me like she's got a life sentence.

"I'll see you on the other side." I steal another kiss.

"Don't forget about me." She nibbles at my jaw.

"That would be impossible."

We part on a final kiss, but after our time together this weekend, I know it's only the beginning.

TWENTY
STELLA

FROM MY BEDROOM WINDOW, I watch Jasper's parents' car pull out of their driveway, then reach for my phone. After days of being Sadie's bitch—I mean helpful bridesmaid—I feel like Rapunzel staring down from her tower looking for an escape. Also, I haven't seen Jasper in two days and I'm going through withdrawals, not to mention the urgent issue that we haven't had sex yet.

> What's up?

Finishing up with some work. What are you up to?

> Casually stalking your house to see when you might be alone.

I'm honored. And alone. My parents went to a movie and Juniper is hanging out with friends downtown.

> I'm coming over.

See you in a minute.

I drop my phone on my bed and run to the bathroom. I use mouthwash, and put deodorant on, spritz some perfume, before giving my makeup and hair a quick once over.

Jasper and I are going to have sex.

I'm more nervous now than I was a few days ago when I thought we'd be having a ton of sex on our getaway. I don't question why, because there's no time for that. Jasper and I need to have sex before the universe gets in the way again.

On my way out of my bedroom, I grab my phone and type out a quick text to Pippa.

> Jasper and I are having sex.

She responds right away.

> Right now? Why are you texting me?!

> No, silly. We're going to. Finally. I'm so nervous! Tell me not to be nervous.

> It's normal. You're nervous because you like him.

> Careful...

It's a teasing warning, but it also makes me stop and think. She's right. I do like Jasper. I told him at the cabin. I might more than like him.

> I need to write a sex scene in this book. Take notes so my characters can live vicariously through you.

I laugh, but also wish my friend could get out from behind her computer.

> I'm happy for you, but also be careful. I don't want to see you get hurt.

> Jasper can't hurt me. I've built up a resistance over the years. I'm rubber, he's glue.

> That's going to be an interesting pairing. Be safe & call me later.

> I will.

My head wants to heed Pippa's warning, but it's not in control of this operation, my body is. And according to my sex-starved body, we are all systems go with Jasper.

Downstairs, I step into my boots and rush out the door, not bothering with a coat since I'm only running across the street.

The late afternoon sun has dropped low in the sky. It's dusk, but the brightness of the snow keeps the light from completely fading. It's my favorite time of the day.

It reminds me of the night Jasper and I spent at the cabin.

When I glance back to Jasper's house, he's there with the front door open, leaning against the door frame.

"Are you coming in?" he asks.

I rush toward him, lunging at him at the last second. He catches me, his hands cupping my ass to lift me up and into his arms.

Wrapping my arms around his neck, I kiss him hard.

"Now that's a greeting." He smiles, one hand wrapping around my waist to support me while the other pulls off my boots from where my feet are latched behind his back. "Can I get you anything? Water? Tea? Wine?"

I shake my head. I'm perfectly hydrated for the occasion.

"Do you want to watch—" he starts.

I cut him off with a kiss, then slowly pull back.

"Take me upstairs, Jasper."

He searches my face for clarity. I nod. *This is it.*

Finally, there's nothing holding us back.

We're a mess of urgent hands and hungry kisses as Jasper climbs the stairs with me in his arms. When we reach his bedroom, he walks us in, then kicks the door shut behind us, before reaching back with a hand to lock it.

Jasper releases me and my legs drop to the ground. His hands make quick work of lifting my sweater up and over my head. I reach for the hem of his sweater to do the same for him.

He unclasps my bra. I unzip his pants. All the while our mouths stay connected.

Then, he's pulling me in close, our chests are skin to skin and nothing has ever felt so good. The tips of my nipples brush against the smattering of hair on Jasper's chest, teasing my sensitive skin. Every nerve ending is alive and I think I'm going to need him to take me right this second.

He must know how impatient I am because he lifts me up and sets me on his bed. My jeans are unbuttoned, and along with my underwear, pulled down my legs in record time.

But Jasper isn't stripping off his pants like I'd hoped he would. Instead, he pulls out a roll of red ribbon.

I smirk. "You just happen to have that lying around?"

"It was left over from wrapping your present." He nods to my ears where the diamond bows are resting.

"What do you want to do with it?" I ask, anticipation building in my core.

He leans over me, sweeps my hair behind my shoulder, and presses his lips against my jaw before he tells me, "I want to tie you up with your hands behind your back, then eat your pussy while you ride my face."

I thought we were just going to have some missionary sex, but it sounds like Jasper has other things in mind.

And I like a man with a plan.

I've never done anything like this before. Never been tied up or bound in any way. The imagery of the scenario Jasper is depicting sends a fresh wave of lust between my thighs. My thighs are soaked and I'm not even thinking about how embarrassing it will be for him to see me this way. Wet and needy for him. Years ago, I would have thought he wanted to tie me up and leave me in an embarrassing situation. Now, I'm not even questioning his motives one bit. I don't know how we got here but I want to see this through. I need it. I have to know what it feels like to have Jasper inside me.

"Are you going to let me play with you, Stella?" Jasper's head dips to suck my nipple into his mouth, flicking it with his tongue, while one finger teases my entrance. "Wrap you up like the gift you are? Just for me."

"*Yes,*" I answer around a moan at his skillful tongue's appreciation for my nipples.

"That's my girl."

I like the sound of those words way too much.

My hips rock forward trying to capture more of his finger, but he's already pulling away.

He tugs several lengths of ribbon from the roll, measuring them out with his arm span, then makes a cut with his teeth. With the ribbon in his hand, he moves in behind me on the bed and settles against the pillows.

"Come here." He crooks his finger and I climb over his lap, pressing my center against the bulge in his boxer briefs. I know there's going to be a wet spot on the material from me, but I don't care. I'm marking my territory.

He pauses a moment, looking intensely into my eyes. "Do you trust me?"

"Yes." The answer comes easily. So easy in fact I wonder how we got here. Me, naked in Jasper's lap, trusting him to tie me up and fuck me senseless.

Leaning forward, Jasper crisscrosses the ribbon under my breasts, then wraps it behind my back and over my shoulders, before bringing it down the center of my body to hug the insides of my breasts. He finishes it off by pulling the ends behind my back.

"Have you done this before?" I ask, suddenly curious if these ribbon-tying skills are seasoned. Which is followed by a sickening churn of my gut from the jealousy that the thought brings.

"No." His answer eases the knot in my belly. "But I'm a quick study."

"So, you Googled this?"

"I was a Boy Scout. We were always tying knots for fun."

"Bet you never imagined using that skill for this scenario."

He cups my jaw with one hand, tilting my chin until our eyes meet. "Stella, there isn't a way I haven't imagined having you. This one included."

Hearing that he's thought about this only makes me more excited to try it. For an odd reason I can't explain, fulfilling Jasper's fantasies just became mine.

Holding my arms in place behind my back, he presses a kiss to my lips before wrapping the ribbon around my wrists and tying a bow. At least that's what it feels like. It's hard to see and when I turn my head and move my arms to look, the ribbon tightens, squeezing my breasts.

The edges of the ribbon are soft, but when I move and it pulls taut, there's a biting edge to it that only heightens my arousal.

Now I'm even more desperate for friction but I'm completely at Jasper's mercy. And he doesn't seem to be in a hurry.

"Look at you, Stell. You look so pretty wrapped up. So delicate and exquisitely tied."

He traces a single finger along the lines of the ribbon, admiring his work. Admiring *me*. I've never felt so worshipped, so beautiful. It's surprising that I can feel this way with Jasper when all we've ever known is arguing and competition. But the last few weeks have shown me a different side of him. Of us.

I watch his gaze follow his finger, satisfaction and adoration filling his gaze. That same finger teases circles around my nipple before pinching the hardened peak.

I cry out at the sensation.

"I should have known you planned to torture me." I'm half teasing but also starting to wonder what I've gotten myself into by giving Jasper so much power. All the power.

"I don't want to torture you, Stella." He slips a hand between my center and his boxer-clad erection that I've been shamelessly rutting myself against. "It's the truth." He dips one finger inside me as he kisses my jaw. "I swear." He pumps a second one in. I gasp and our eyes lock as he finger fucks me, adding a third. "Scout's honor."

He uses his free hand to pull on the ribbon behind me,

which only intensifies the sensation between my legs. His mouth latches onto the sensitive spot between my neck and shoulder, licking and teasing the flesh there.

When his thumb joins the party, I'm a goner.

"Jasper," I moan.

"Yeah, baby?" He says it so tenderly it makes my chest squeeze.

"I'm going to come."

"That was the plan."

Another thrust of his fingers and I'm spiraling out. My muscles clamping down so hard around his fingers it's almost painful. But the pleasure is worth it.

The aftershocks of my orgasm are still pulsing through my body when Jasper grips my hips and lifts me up his torso and over his shoulders until I'm positioned above his face.

It's a vulnerable position to be in. My center hovering over his mouth and chin. My whole body feels like a holiday Jell-O mold. Jiggly and barely set.

"Jas, I've got nothing to hold onto."

"I've got you." He wraps his arms around my ass, holding me firmly in place above his face.

I can feel my wetness leaking out between my thighs.

"I'm so wet."

He licks his tongue through me. "Yeah, you are. I fucking love it."

I'm sensitive from my orgasm, but Jasper doesn't hold back. He's there, tongue warm and insistent as he licks my center.

"Grind that pussy on my mouth, Stella. I want you to fucking drown me."

So, for once, I give Jasper what he wants, and I smother him with my pussy. It's mutual satisfaction as he licks and sucks and fucks me with his tongue. It only takes a minute and I'm lost in the sensation of him again, chasing another orgasm until I'm boneless and collapsing in his arms.

TWENTY-ONE
JASPER

ONCE STELLA STOPS PULSING against my tongue, I lift her off me and sit up so she's cradled in my arms.

Gently, I untie the ribbon around her wrists and start to unwrap her.

"Is that what you wanted?" she asks. "To unwrap me like a present?"

"Yes." I swallow thickly because seeing Stella wild and raw, with no guard up between us, is exactly what I wanted for Christmas. Unwrapping her, and knowing she's mine is the ultimate gift.

There are a few pink streaks under her breasts and around her waist from the ribbon. I press my lips to them, and she shudders, gooseflesh breaking out across her skin.

"Are you okay?" I ask.

"I'm more than okay." She laughs, light and airy.

I pull the covers open, then roll Stella onto her back to tuck her inside. She's drowsy and delirious, and absolutely stunning with her pink cheeks and wild hair.

"I might upgrade you to *real* boyfriend after that."

Her eyes fly open and her mouth parts like she's alarmed at having said that out loud.

"It's okay, Stell. I won't tell anyone you're madly in love with me. It can be our secret."

Her face attempts to scrunch with mock distaste but she's either too relaxed to put up a fight or she knows it's useless. Either way, I kiss her sweet and slow so she knows that's exactly how I'm going to fuck her.

"Take off your boxers, Jasper."

I remove the last item of clothing between us and Stella reaches down to stroke me. I cup her breast, squeezing and kneading her softness. I can't get enough of her curves. The feel of her skin, so soft and plush beneath my palms. She fucking owns me.

Our kiss turns molten and Stella lines me up with her entrance, nothing between us.

"Jasper. I need you. Please." She arches her hips upward, pressing my tip inside.

"Should I get a condom?" I ask as her slick heat wraps around the crown of my cock.

Fuck. I'm barely inside her and I already know she's going to ruin me. I want Stella bare and raw. I don't want anything between us.

She shakes her head.

For a split second I imagine burying myself so deep in Stella, filling her with my cum, and putting a baby inside her. It's not a rational thought. We're not there yet, but it's what I want for our future. Seeing Stella pregnant with our child.

"I'm on birth control." She snaps me out of my foolish thoughts. "I've been tested. I'm good."

"Same," I confirm.

Holding her gaze, I rock my hips forward, pushing all the way inside her.

For a moment, I hold myself still, indulging in the feeling of her stretched around me. Her muscles contract, then relax as she adjusts to my size.

I start to give her short strokes, slowly working myself in and out until my cock is fully coated in her wetness and she can take all of me at once.

"Jasper." She clings to me while I pump into her over and over. Thrusting my hips in tempo with her soft moans. "God." She sighs. "It's so good."

"I know, baby. We fit together perfectly."

Being inside Stella is everything and nothing I could have imagined.

It's the culmination of every feeling I've had over the last two decades.

With every stroke, I'm telling her how I feel.

I've waited for you.

I want this with you.

I love you.

I hold myself deep inside her, giving her measured strokes, as the soft patch of hair between her thighs tickles the base of my cock. A tingling sensation starts to build at the base of my spine.

Stella's face says it all. Pleasure. Connection. *Trust.* This thing between us is as deep as I'm buried inside her.

"Come on, Stell. I need to feel you come. I need you to milk every inch of me."

I thrust into her again. This time her hands drop to my ass to hold me still inside her. A moment later, she spasms around me, squeezing my cock and pulling me impossibly deeper. Relief washes over me as I stroke *one, two, three* more times before I spill inside her.

I pull back to look at her and notice her eyes are glassy.

I hold her face in my hands. "Don't cry, Stell."

"I'm not," she says defiantly, all while tilting her head up toward the ceiling to keep the tears at her lash line from dropping. "It's hormones, and the holidays, and I don't know...everything."

"I know." Because I know exactly how she's feeling. I've had these feelings for much longer. What I've had years to process is hitting Stella like a tidal wave, and all I want to do is hold her until she can find her equilibrium again.

I kiss her softly, then wrap her in my arms. After a while our kisses become deeper, then Stella reaches between us, stroking me until I'm rigid and the frenzy of having each other starts over again.

TWENTY-TWO
STELLA

I SLIP out of Jasper's reach and grab his t-shirt from the floor to pull over my head.

"Come back here. I promise I won't touch you. I only want to cuddle."

His eyes are innocent, but the monster erection between his legs is telling me another story.

"You think I'm falling for that?" I scoff.

"It's true," he says innocently before adding a wicked smile. "At least for the next ten minutes."

But I need a moment to collect my thoughts. The intensity of having sex with Jasper for the first time was more than I anticipated. The second time was even more powerful because that's when I realized, just like our first kiss, it wasn't a fluke.

"How much stuff do you have in here? It's like a museum for the early to mid-two-thousands."

"My parents haven't gotten rid of anything." He props himself up on one elbow, watching me peruse his closet.

"Our style was horrible."

"I think I was a good dresser," Jasper winks at me.

"Yeah, you did look good but I never would have admitted it."

"And why's that?" he prods.

"Because we were sworn enemies and a compliment was against the rules."

"I think we need to amend the rules."

"Why's that?" I ask, my eyes tracking his naked body as he pulls on his boxer briefs and moves to stand behind me.

He pulls my hair back and presses a kiss to my neck.

"Because I'd like to compliment you on how fucking amazing your pussy feels wrapped around me."

An airy laugh escapes me but it catches in my throat as I notice a familiar scene peeking out from between Jasper's vintage clothing. I push the hangers aside to reveal it.

There, tucked in the back of his closet, behind a glass frame, is my winter wonderland drawing.

TWENTY-THREE
JASPER

I HOLD my breath as Stella takes the framed drawing off the wall to examine it.

I know what she's looking for. Evidence of its destruction. Ripped pieces, torn edges. Because that's what I made her believe I'd done to it.

I remember the moment like it was yesterday. The look on her face and how it had given me a rush of satisfaction, and then, for the first time since we'd been trading jabs, an overwhelming sense of regret.

I'd felt like shit for hurting her like that.

Making her believe I'd destroyed something she'd worked so hard on. That I'd purchased her drawing with the intention of hurting her, when in fact I bought it because I knew it was special. She was special. And I wanted a piece of her for myself.

I never told Stella that I didn't tear up her drawing. I didn't think she'd believe me.

But that day, the look on her face, and the heartbreak in her eyes, was the moment everything changed for me.

Yes, I kept playing the game with her. Kept competing

and keeping tabs on her. But it wasn't with the hope of crushing her spirit, but the hope that she would get to know me and we could be friends. More than friends.

Slowly, she turns around to face me.

"What is this?" Her voice is barely above a whisper as she shakes the frame. "Did you make a copy?"

"No." I swallow. "That's the original."

"You didn't rip it up?" she asks in disbelief, her eyes turning glassy again.

I shake my head. "No."

I don't know what to say. I should have told her. I should have given it back to her at some point as a peace offering. There are a thousand things I wish I would have done differently with Stella, but then we might not have gotten to where we are now. Or at least where we were twenty minutes ago when she shattered around me and I saw the potential for forever with her. And fuck, until this moment where my stomach is filling with dread, I loved that we made it here.

Now, I'm searching for the right thing to say, uncertain how to navigate this. I watch Stella's face go through all the emotions. The shock of seeing her drawing again, the confusion that it was never ripped up, and then the hurt that I've had it all this time and never told her.

A single tear snakes its way down her cheek. "Why didn't you tell me? Why did you let me believe—" She chokes back a sob.

"I—I didn't know if you'd even believe me."

"God, Jasper." Her voice quivers with hurt. "That was the moment I started hating you."

Her words pierce my heart, but I have to keep going. I need to tell her the truth now.

"It was the moment I started loving you. I didn't know it

at the time. It took me years to figure out what that feeling was. And then, it felt like it was too late."

"What did you say?" She wipes furiously at her tears.

"I love you, Stella. I've loved you all this time."

"That doesn't make sense. You hate me. Or you did." She shakes her head, trying to reconcile my admission with our past rivalry.

"You told everyone I was joining a convent after graduation."

"I was an idiot. I thought that would deter other guys."

"Well, it worked." Her face flushes red with anger. "I went to prom alone."

"I wanted you to myself."

"But you never had me, Jasper. You never told me how you felt. You just made sure I was alone and miserable!"

"I didn't know how to tell you back then. How to make you trust me after everything. I thought that this arrangement would help—"

"You lied to me. You didn't need a fake girlfriend. This was all a setup, but for what?"

"No. Not a setup. Fuck. This isn't how this was supposed to go."

"How was it supposed to go, Jasper? Was I supposed to be happy that you lied to me all these years?"

She shakes her head in a swift arc.

"This was a mistake." Still clutching the frame with her drawing, she moves to gather her clothing that's strewn about my bedroom. "A huge fucking mistake!"

"Stella. I'm so fucking sorry." I reach for her, but she pulls back.

"No! I was devastated about this drawing. What I thought had happened to it. That you would be so cruel to do that. And then, over the years, there would be moments where I thought you might not be the jerk I thought you were, then some other fight or competition would flare between us and my walls would go back up.

"Stay away from me, Jasper. Don't call me. Don't text me. And do not come to Sadie's wedding tomorrow."

She slams my bedroom door, and I stand there stunned.

How did this go from the best night of my life to the worst so quickly?

Fuck. I have to fix this.

Grabbing my jeans off the floor, I hop on one foot, then the other, trying to pull them on as fast as I can without tripping. With no time to waste, I rush down the stairs to see if I can catch her, but the house is empty. She's already gone.

I don't bother with shoes. I throw open the front door and run after her.

Shirtless, with my jeans unbuttoned and bare feet, I run out into the cold evening air.

"Stella!" I yell, like the lovesick fool I am.

She turns around, fury in her eyes that I haven't done what she asked.

"Leave me alone, Jasper."

I'm halfway across the street when my feet go numb from the snow underfoot, but I keep moving.

"I'm not done talking, Stell."

"I don't want to talk to you. Ever again!" she screams. I'm sure the neighbors are all peeking out their windows right now. I don't care. What we found these past ten days, the chemistry, the passion, the tenderness, is something I'm not willing to give up on no matter how hard Stella pushes back.

"That's too bad, because I need to tell you something," I say, continuing my charge across the street and onto her lawn.

"Jasper," Stella warns.

"Stella," I counter with the same obstinacy. "This hasn't been one-sided. You've slung plenty of mud yourself. And I think you like to fight with me because it allows you to bury how you truly feel."

"And how is that?" she asks in a mocking tone.

"I love you, Stella. And if you look past all the petty bullshit that's happened between us, you'd realize that you love me, too."

She sets the frame down carefully on one of the porch chairs, then drops her clothes on top of it. She's wearing only her fuzzy pull-on snow boots and my t-shirt.

At first, I think she's rushing toward me, but she stops short on the sidewalk and bends down to the snow-covered lawn. Scooping snow into her hand, she forms a ball and sends it flying toward my head.

At the last second, I duck.

I don't know what I expected, but an impromptu snowball fight wasn't it.

"You missed," I call. Should I be engaging her this way? Probably not. But what I've learned about Stella is as long as she's willing to stay and fight, there's still hope. It's when she shuts down and disengages that I'm at a loss.

I'm too busy gloating about her miss that I don't see the next one coming. It hits me right in my chest. My bare chest.

Not only is the snowball perfectly packed so it doesn't completely break when it hits me, but my lack of clothing has my bare skin absorbing every icy edge of it.

"God damn it, Stella." I rub my aching chest. "That hurt."

"Good," she retorts, reaching for more snow.

She launches another snowball in my direction, but this time I'm ready for it and duck.

Shielding my face with my hands, I inch closer to her.

"Can you stop throwing snowballs so we can talk?"

"No." She reaches toward the ground to reload but knowing she's not armed yet, I make my move.

Rushing toward her, I scoop her up into my arms. She kicks and flails like a greased eel, trying to squirm out of my grip.

She slams her boot into my shin and I release her as I start to fall forward. If I had ever been concerned about Stella living in the city by herself, it's clear from her self-defense moves I've got nothing to worry about.

At the last minute, I reach for her to pull her close so she doesn't fall on the ground.

We fall into a heap against the snow, my bare back against the frosty ground while Stella's body crashes against my chest. I should feel the cold against my skin, but nothing matters because all I can think about right now is making this right with Stella.

Her hands, icy from gathering snow, press against my chest so I move to cover them with mine, for warmth and the hope that I can keep her with me long enough to explain things.

"Please, Stell." I squeeze her hand.

I've seen what we could be, the moments where she lets her guard down with me and it's fucking magic. I want it. I want her. For the rest of our lives.

But what if she can't forgive me?

Tears spring to my eyes, the emotion hitting me so intensely that this could be it. She could be done with me but I'll never get over her.

She takes a shuddering breath, her features softening as she stares down at me in the snow.

"I never wanted to hurt you. I was stupid and immature. I know that doesn't excuse my behavior, but it's the truth. I love you, Stella. I always have."

She shakes her head, huge tears brimming at her lash line.

"I hate you, Jasper," she whispers, but there's no anger behind her words, only sadness. "I hate you for making me want you, then hurting me all over again."

I release her wrists, and she sits back on her heels, putting distance between us.

I sit up, still holding her gaze with mine.

"Please, Stell. I'm sorry. I don't want to fight with you. I never did."

Without another word, she sits up and brushes off the snow clinging to her legs. She returns to her porch, gathers up her things and goes inside.

I sit there for a minute longer. I don't feel the cold anymore. Everything is numb.

Eventually, I stand up, and make my way back across the street to my house.

When Mrs. Peterson from down the street passes me with her Corgi, Wilson, I don't even have the energy to be embarrassed by my appearance, and she must have the wherewithal to know not to ask.

Back inside the house, Juniper is in the living room watching television.

"What happened to you?" she asks, her eyes catching on the huge red spot where Stella nailed me in the chest with a snowball.

"Stella St. James."

She gives me a tight smile. Enough said.

TWENTY-FOUR
STELLA

"JASPER'S not coming to your wedding," I fume the second I walk in the house.

I hold onto my anger, because if I let myself be sad, then I'll rush back across the street and cry in Jasper's arms. The way he looked at me. The way he was gutted when I said I hated him.

I'd hoped it would make me feel better, but it did nothing to soothe the ache in my chest.

"Are you serious?" Sadie exclaims, hands waving near her head. "Oh, god. We're going to have to redo the seating chart."

I collapse onto the couch.

I'm wet and freezing in Jasper's t-shirt from our fight in the snow, so I pull a blanket off the back of the couch and wrap it around myself.

Should I care about a fourteen-year-old drawing that isn't even as good as I remember it being?

Probably not.

But it's not about the drawing. It's how cruel Jasper had

been. And how his opinion, whether I wanted it to affect me or not, shaped how I saw myself as an artist. How at twelve years old, a time in life where one's sense of self-worth is fragile and taking shape, yet one's self-esteem is most often laid in the hands of peers that are struggling to find their own sense of self.

That moment with Jasper and my drawing, willing or not, is a core memory.

Jasper, bare and gently pressing himself inside me for the first time, is, too.

Fuck. How did everything go sideways so fast?

More tears cascade down my cheeks.

"Oh, Stella." Sadie throws her arms around me. "It's okay, we'll figure out the seating chart."

"It's not that," I sob. "I mean this in the nicest way, Sadie, but I don't give a fuck about the seating chart. I just had a fight with Jasper. And we had sex. It was unbelievable. Like life altering because not only did I come but I think I might be in love with him but then he hurt me."

"What?!" She rears back, eyes ablaze with rage. "Are you serious? Where? How? I will fucking murder him. They'll never find his body."

"No, he didn't physically hurt me. Emotionally." I press a hand to my chest. "My heart."

She wraps me in her arms and rocks me soothingly while I cry.

Jasper framing my drawing. Finding out it was never destroyed and he had it all these years. His feelings for me and the fact that he was never really competing but only wanting me to see him. And I did see him. There were moments over the years where I wondered if there was something more between us but then we'd be right back to fighting and I'd feel silly for letting myself think that.

Because the truth is, over the last twenty years, through our rivalry and competitions, I'd slowly given him pieces of myself.

And tonight, I'd given him everything, only to be left with a bruised heart.

TWENTY-FIVE
STELLA

SADIE AND TOM'S wedding is pure magic. Nestled in a rustic lodge surrounded by snow-covered pine trees overlooking Snowcap Lake, the reception tables and dance floor are draped in a canopy of twinkling lights and crystal chandeliers that look like dripping icicles. The tables are adorned with vintage candle holders and centerpiece arrangements of sugared berries, winter florals, and succulents.

I lift the frosted lowball glass I'm sipping cognac from and let the notes of vanilla, spice, and caramel dance on my tongue before burning their way down my throat. I don't even know where I got this drink from. One of Tom's parents' friends was drinking it and suggested I give it a try. I figured I had nothing to lose at this point.

With Jasper not in attendance, Sadie said fuck it, kept the seating chart as it was, and left his seat empty. I know she thought he would show. Make some grand gesture and declaration of love, but I reminded her that he already did that on our front lawn and I pelted him with snowballs in return.

I brought Gideon, the Christmas gnome that Jasper bought me at the tree farm craft market, as my emotional support gnome and sat him on the table beside me. When I sense people attempting to come up to talk to me, I start feeding Gideon cake and that seems to keep them from stopping.

"Hey, Stella."

I look up to find Daniel standing at the table. I guess Gideon didn't do the trick.

"I'm here to check on you."

"That's nice," I say, noticing my words are starting to blend together, so I push the cognac glass away. "Thank you."

"Full disclosure. Sadie asked me to. I'm with Cady now, so I don't want you to get the wrong idea."

I have to laugh at the irony in his statement.

"Oh, right. You and Cady. That's great."

"I know it seems sudden, but when you know, you know."

My thoughts wander to Jasper. Or you think you know and everything is completely different than what you thought.

"And no disrespect to what we had." He motions between us. "It was one night of passion, but me and Cady are something else. I can't even describe it."

Daniel makes this relationship thing sound so easy. I wonder if I'm missing something.

"Can you try?" I ask, genuinely curious.

Since our fight in the snow yesterday, Jasper hasn't texted or called. I didn't expect him to, but it does feel like after all these years, and everything we've been through, he wouldn't so easily back off. I know I'm not being fair. Wanting my cake and eating it, too, and all that. But after a night of restless sleep, and a day filled with watching Sadie and Tom love each other for exactly who they are, faults and all, I'm starting to rethink things.

Maybe coming clean to Daniel will make me feel better about the situation. It can't hurt.

"Jasper and I were never really together."

Daniel blinks. "What?"

"I saw you at the airport and didn't want to deal with your flirting so I pretended Jasper was my boyfriend."

His mouth starts to curve into a smile and then he laughs. "That's a good one."

"I'm serious."

"Nah." He shakes his head in disbelief. "You and Jasper are relationship goals. I knew from watching you with him I had no chance. It's the best thing that could have happened because seeing you two together helped me get over you and find Cady."

I sigh. If Daniel won't believe that Jasper and I weren't a real couple even when I tell him to his face, then I'm at a loss. "Okay. Sure. We're madly in love. But we have some unresolved issues and now we're not together anymore."

"You two are going to work it out. I know because I see the love between you." He leaves me with that, rushing off to find Cady in the crowd of guests on the dance floor.

God damn it. Even Daniel was thoroughly convinced we were a couple. I'm the one who couldn't get it through her thick skull.

But that's not true.

I suspected it on Christmas Eve. When Jasper gave me the bow earrings. It wasn't the value of them, but the fact that he paid attention to something he knew I'd like. And I discovered there was no setup by his mom, and there never had been. When we slow-danced in his living room and he kissed me like I was his. There was an inkling then.

And then in Jasper's room after he had me for the first time, I knew what it was.

Love.

That hadn't been the scary part.

It was seeing my drawing and hearing that Jasper had loved me, not for the past ten days, but the last ten years, even more than that. That moment had hurt because it felt like a betrayal. I'd built the foundation of who I was largely on my rivalry with Jasper. Hearing that he was never playing the game the way I was made me feel silly. I'd put trust in what we were, but that no longer existed. It felt like a race that only I was running while Jasper was already at the finish line waiting for me to catch up.

It wasn't our relationship that was fake. It was our rivalry.

And I can hold onto the way I felt about him all those years ago or I can move forward with how I feel about him now.

Through the blur of my unshed tears, I scan the room, my eyes are desperately searching for...for what? Jasper isn't here.

My chest aches with the thought of him sitting at home alone. On New Year's Eve. He should be with the one he loves. He should be with *me.*

But I'm here and he's not.

Then, it dawns on me that I can go find him. All of my bridesmaid duties are done and the last event of the evening is the countdown to midnight.

"As we get closer to counting down to the New Year, Tom and Sadie ask that their guests gather by the large windows overlooking the lake for a fireworks display accompanying the new year's arrival."

Glancing out the window, my heart drops.

It's snowing. Jasper's house is miles away and I'm in heels with no transportation.

I'll borrow someone's car.

But then I remember all the guests were shuttled here from the parking lot down the mountain. Even if I can borrow someone's car, I'll need to ride the shuttle first. And there's the small issue of the cognac I've been drinking and the fact that I shouldn't be driving anywhere.

"One minute until the new year," the DJ announces.

My heart races at the announcement. I'm stuck. There's no way to make it to Jasper.

I pull my phone out to call him. Even if I can't see him, maybe I can talk to him. Maybe I can make things right before the new year.

His line rings and rings, but he doesn't pick up.

Defeated, I shove my phone into my clutch.

As the countdown draws closer to midnight, the sound of the crowd chanting is drowned out by the whooshing in my ears. My throat tightens and fresh tears brim at my lash line. All I want is to see Jasper. To kiss him at midnight and tell him I love him.

Five...

I let the tears fall. There's no point in holding them in anymore.

Four...

I start moving toward the exit. I can't be here right now. I need fresh air.

Three...

Just as I'm about to reach the door, a firm hand grips my

elbow and in the next second I'm spinning in my heels. Around, around, until two steady hands grip my waist, catching me as the momentum threatens to spin me past my destination.

Two...

Standing in front of me, in jeans and a wool coat with a dusting of snow in his hair, looking handsome as hell, is Jasper.

One...

I open my mouth to speak, but my words are swallowed by Jasper's demanding mouth on mine. My hands find purchase in his thick locks which are damp from the snow. My whole body sighs with relief and then I hold on for dear life as Jasper and I work out all the emotions of the last twenty years in this one kiss.

When our mouths finally release each other, I'm breathless and weak in the knees. It's a good thing Jasper is holding me up.

"You came." My hands slide inside his coat to feel his warmth.

He shakes his head. "I tried to do what you asked but I couldn't stay away, Stella. You said it was bad luck to start the year with someone you hate, well I couldn't stand the thought of starting the new year without the person I love."

He brushes his thumb against my wet cheek. "I love you, Stella. Even if you can't move on from the past and trust that my feelings are genuine, I need you to know that I'm sincere."

My hand covers his, holding it to my face, and the sound that comes out is half-sob, half-laugh.

"Don't cry, Sparky." He says it with so much affection, I think I'm going to melt into a puddle.

"They're happy tears." I shake my head, laughing again. "I'm happy you're here and that you still love me."

He shakes his head, a low chuckle releasing from his chest. "It's been one day."

"I know, but those snowballs were pretty hard." Against his sweater, I rub my hand over his chest where I pelted him, then drop my lips there for an apologetic kiss. "I'm sorry."

"I'm sorry, too." His lips are on mine again and I could easily get lost in his taste.

But then I remember, I haven't even told him the most important part. I part our lips to lock my eyes on his.

"I don't hate you, Jasper. I don't think I ever did and that was the most startling realization. It's why I ran, because it was scary to find out that I'm in love with my nemesis."

His hazel eyes widen like he can't quite believe what he's hearing.

"What are you going to do about it?"

He wraps his arms around me. His hands sliding down my back to cup my ass.

"I'm going to kiss you again." I press my lips against his softly. "Take you home with me tonight. And when you move to New York next month, make you move in with me."

"I love this plan." His lips tease along my jaw and I know I could easily get lost in him again, but I still haven't officially said it.

My hands cup his face so I can pull him back to meet his eyes again.

"I love you, Jasper."

His eyes light up like a kid on Christmas morning.

"Say it again."

"I love y—" Jasper doesn't let me finish. His lips capture mine and we're lost in each other all over again.

TWENTY-SIX

JASPER

"NO ONE'S AT MY HOUSE." Stella pulls my hand, urging me across the street. "Sadie and Tom are at the hotel. Daniel's with Cady, and my parents, of all people, went downtown with some of the wedding guests."

It feels like we're teenagers sneaking around.

She opens the door to her house, then guides me upstairs, and I wonder if this is what it would have been like if we hadn't been fighting all those years.

I lie on the bed waiting for Stella as she instructed, and try to imagine what her room looked like when we were younger. Currently it's light gray walls and a white duvet. It's exactly as she described—guest room number one. I can't wait to see her place in New York. I can't believe she asked me to move in with her. That might have been the Cognac talking but I'm hopeful she was serious, because I want to wake up next to her every day and fall asleep next to her every night.

Stella comes out of the bathroom. She's wearing a fuzzy blue bath robe, and her face is now free of makeup. With no preamble, she crawls over me and straddles my lap.

"I love you," she says, unbuckling my belt, and a moment later she's got me in the palm of her hand. It's only a few strokes before I'm impossibly hard and leaking at my tip.

I cup her jaw and pull her close. "And I love you."

She parts her robe, and then I'm sinking inside her.

Her mouth parts on a gasp, and I groan at the feel of her tight walls clamping down around me.

Stella St. James loves me and she's riding my cock so fucking well, I can't believe my luck.

It doesn't take much to send us both over the edge, and once I've cleaned her up, Stella stands at her window, looking out across the street at my parents' house. I step in behind her, wrapping my arms around her and snuggling into her fuzzy robe.

"You know how much this pains me to say it, but your house is the official winner of the Whistler Lane holiday lights display contest this year."

I smile against her neck.

"Don't rub it in. You sabotaged my house with mediocre light installation."

"It's not a contest."

"Actually, that's exactly what it is."

"Not for me."

"What do you mean?" she asks, her eyebrows melting together in confusion. "It's always been a contest. You've always put up an impressive display because you wanted it to be better than mine."

"No. I knew how much you love Christmas lights so I wanted you to have the best display to look at." I motion across the street to my house. "The light display is one thing, but really, anything I did, it was always for you, Stell."

She turns to me and wraps her arms around my neck,

her expression one of contemplation. "I feel like an idiot for not noticing what was between us."

I shake my head. "You shouldn't. I was good at making sure you didn't know. It was my armor."

She presses a soft kiss to my lips, then one to my jaw. "Well now I need to show you just how much I love you."

"What did you have in mind, Sparky?" I ask, my hands dipping inside her robe to fondle her ass.

"I think you know, Snowflake." She turns and pulls me back toward the bed where I peel off her robe.

There, under the glow of the Christmas lights shining through the window, we sink into each other. We let go of the past, and all the baggage from our old rivalry to ring in the new year exactly how I've always envisioned, in each other's arms.

EPILOGUE

ONE YEAR LATER

"What do you say, Stell? You up for a gift wrap contest?" Jasper spins a roll of ribbon around his finger suggestively. I'm always up for a friendly competition, but it's fun to mess with him anyways.

"Are you sure? I wouldn't want to bruise your already damaged ego from my Holiday Pictionary victory last night."

Last night, after Jasper and I had a romantic dinner downtown then walked around looking at shops and Christmas light displays, I rallied Jasper's mom, his cousin, Milo, his Aunt Melanie and Uncle Ron, and Juniper to a win. I made a point of having his Uncle Ron on my team again, I needed to prove that I could win with him, even after he was two spiked eggnogs in.

He laughs. "You definitely improved upon last year's showing."

It's our first Christmas officially together and we're

making the most of it by doing all the holiday traditions of his family, and mine.

We're staying with my parents. Sadie and Tom are spending Christmas with Tom's family this year, and we'll see them in a few days for a post-Christmas celebration.

Jasper and I have also been creating a few new traditions of our own. Like the annual ornament exchange we've decided to have, both of us making or gifting the other an ornament that represents a memory from the past year. This year Jasper gave me a photo ornament, one side has a photo of us standing next to each other, not voluntarily, in third grade while the other side has a picture of us from February when we moved into our new apartment together. On the bottom, the engravement reads: *Our love was worth the wait. Love, Jasper*

The ornament I gave him was a ceramic snowflake, inspired by the one he had made me all those years ago, hand-painted and glazed at the local pottery painting shop in Cedar Hollow. On the back I wrote: *From snowball fights to cozy nights. Love, Stella.*

Following with our new traditions, we're back at Toys for Tiny Hearts volunteering our wrapping skills for the charity and Jasper is foolishly challenging me to a wrapping contest. Will this man ever learn?

He gives me a knowing smile. "I've been preparing for this moment all year."

My brows lift. "Ah, so you're saying this might actually be competitive, unlike when I whooped your a—"

He cuts me off with a kiss, but a moment later, I can feel his body tense against mine. His hand trembles against my skin where he's cupping my jaw.

"Hey, you okay?" I ask, reaching up to place my hand over his.

"Yeah," he chuckles, flexing his hand before running it through my hair, "must be nerves for this intense gift-wrapping competition."

I smile up at him, taking in his earnest hazel eyes and the recently grown scruff on his jaw that I love to feel against my thighs. "You're cute when you're nervous."

"I thought I was cute all the time?" he quips.

"You are, but you're looking extra adorable in this sweater." I run my hand along the sleeve of Jasper's sweater that is striped in white and red to look like a candy cane. The front has a tree crocheted on it and embellished with beads as ornaments.

I'm one to talk. I'm wearing the sweater his grandmother crocheted for me this year. It's green with white pine trees around the collar. It's so hideous, it's actually really cute.

"Back at you, Sparky."

"I accept your challenge." I brush my lips against Jasper's one more time, then walk toward my wrapping station, swaying my hips flirtatiously because I know he's still watching.

When I get there, Sandy is already rolling over a cart of wrapping paper.

"We got some new designs, Stella. I think you're going to love them," she says, clasping her hands excitedly.

Part of Jasper's plan for Toys for Tiny Hearts is giving platforms to children and teens that are in an art program the charity recently started across the country. This year, kids submitted holiday themed art and a few were selected to have their designs made into wrapping paper for the charity, which is also now being sold wholesale to benefit the art program.

"I recommend you start with this one."

Sandy hands me the roll. I place it into the wrapping paper holder and pull the edge to expose the design. There, printed on the glossy white wrapping paper, are smaller versions of my winter wonderland drawing from seventh grade. It's the coolest thing I've ever seen. My misty eyes search the warehouse for Jasper, but he's not there.

Where did he go?

Then, I turn around and he's behind me. On one knee.

"Jasper." A hush falls over the warehouse, but the faint sound of Bing Crosby's "White Christmas" can be heard playing in the background.

"I've loved you for so many Christmases, Stell, but now, I want them all." He opens the small velvet box in his hand. "Will you marry me?"

I'm nodding my head and crying because of course I want to marry him. He's my best friend and biggest supporter. And I'm the same for him. We challenge one another and give each other space to make mistakes and learn from them. It's a loving adult relationship formed out of our childhood rivalry.

"Yes. Of course." He stands to wrap me in his arms, then pulls back to place the large diamond on my finger.

"Holy shit, Jas." I gawk at the diamond on my finger. "It's huge."

"That's kind of my theme." He gives me a playful wink, and all I can do is laugh and shake my head.

"And the wrapping paper of my drawing is beautiful. I can't believe you did that for me."

"I'd do anything for you." He wraps me up in his arms and kisses me while everyone around us cheers.

"Wait just a second." I walk over to the table and grab the small velvet box from my coat. "I got you something, too."

I pop open the box to reveal the platinum band I had made custom for Jasper. The engraving on the inside, *Stella + Jasper*.

His face lights up like the Rockefeller Christmas tree, eyes bright and dancing with emotion.

"You were going to ask me to marry you?" he asks.

"Yeah, and you beat me to it," I mock pout.

He smiles and pulls out the band, twisting it to read the engravement before glancing down at me thoughtfully.

"It's not a competition, Stell."

A soft laugh escapes me. "Not with you, but I need to lock this down before other women catch on to how great you are."

His lips twitch in amusement.

"I'd say there's no rush. After all, I've already waited twenty years for you."

I playfully swat at his arm, but he pulls me in against him and kisses me, then drops his mouth to my ear.

"Yes. That's my answer. To you. To us. To forever."

THANK YOU

Dear Reader,

Thank you for taking the time to read my book. I hope you enjoyed Stella and Jasper! There are so many books to choose from, so thank you for spending your precious time reading mine. If you have a minute, please consider leaving a review for Hostile for the Holidays. Reviews help indie authors so much!

XO, Erin

ACKNOWLEDGMENTS

Thank you to my family, Eric and my children, for supporting me always and giving me space when the words are flowing.

Thank you to Shayna for helping me create this fun and festive cover! I appreciate your creativity, your friendship, and our coffee walks.

Thank you to my copyeditor, Chelly. It's always a pleasure working with you.

Thank you to my assistant, Taylor, for cheering me on and encouraging this project from the beginning. Your excitement kept me moving forward! And for making the prettiest graphics and keeping everything running smoothly behind the scenes.

Thank you to Jenny at penpal PR for taking on this project last minute.

ABOUT THE AUTHOR

Erin Hawkins is a spicy romcom author who lives in Colorado with her husband and three young children. She enjoys reading, working out, spending time in the mountains and with her family, reality TV, and brunch that lasts all day.